Zero to AI

Business Strategy for an AI-Native World

9 Shifts. 3 Engines. 6 Moves.
Build Your AI-Native Advantage.

Sudhir Kadam

ISBN: 979-8-9986805-2-6 (Paperback)

First Released March 2026

For permissions or inquiries: zero-to-ai@fyda.net

We limit ourselves when
we mistake a constraint
for a law of nature.

Disclaimer

This book is intended as a general reference and for informational purposes only; the author and publisher are not providing legal, financial, or professional advice. The business examples are drawn from publicly available sources and represent the author's interpretation of that information. Some examples have been anonymized or presented as composites to protect confidentiality. The author and publisher disclaim any liability for errors, omissions, or consequences arising from the use of this book.

Author's Note

Why This Book Exists

This is not a book about AI technology. You won't find explanations of neural networks or debates about artificial general intelligence here. Those are important conversations, but they are happening elsewhere.

This book began with a simple question that haunted me: *Why do most companies treat AI like a tool, when it's clearly becoming a new environment?*

We often use new technology to do old things faster. The real breakthrough comes when we stop asking, *How does this improve my current process?* and start asking, *What can this make possible?*

It's the difference between using a smartphone as a better touch phone versus using it as a portal to the world's knowledge, brought in an instant.

We're in that moment with AI.

For past several years, I've worked at the intersection of artificial intelligence and business strategy—advising startups and studying the patterns of companies that weren't just using AI but being transformed by it. A pattern emerged that disturbed me. I stopped looking at AI as technology and started seeing it as **new physics**.

This book is about how AI unlocks new capabilities to rebuild your company around them. It's about business model reinvention. It's for leaders who realize that adding AI to an old business model is like putting a Formula 1 engine on a lawnmower. It may work, but it's a waste of the engine's true potential. Besides, the lawnmower not designed for that force will soon shatter.

The companies winning today aren't those with the best "AI strategy." They're the ones who understood that **the constraints that shaped business for a century have dissolved**. You're playing a different game entirely. This book is my attempt to map the rules of that new game.

A Word on Urgency

Some will read this book and think, "This is too radical. Our industry moves slowly." But the fact is that the distance between "experimenting with AI" and "being disrupted by it" is no longer measured in decades. It's measured in product cycles.

The companies featured in this book aren't just tech startups. They're established players in traditional industries who realized that waiting was more dangerous than moving. They're not future-proofing; they're present-proofing.

How to Read This Book

This isn't a theoretical exploration—it's a field manual for transformation. Each chapter builds toward a complete architecture for AI-native business:

Part I reveals the **New Physics**: the fundamental shifts in capability that change what's possible

Part II provides the **New Foundations**: how to redesign your business model around these capabilities

Part III maps the **Path to Execution**: how to build momentum and avoid the pitfalls that derail most transformations

Read this with your business in mind, not with AI in mind. I don't have all the answers. No one does in a landscape moving this fast. What I offer is something more valuable: a **lens** to see the new terrain, a **language** to describe what you're seeing, and a set of **building blocks** to construct your advantage on it.

My hope is that this book gives you that clarity. Not as an abstract insight, but as a concrete advantage. The ground has already shifted. The question is whether we'll build shelters on the old terrain or foundations on the new.

Let's begin building.

Sudhir Kadam

Contents

Part-I: THE NEW PHYSICS

Every era of business runs on an invisible set of rules. Rules about what is scarce, what is expensive, and what can and cannot be done at scale.

For decades, reasoning lived in people. Experience accumulated slowly. Decisions were localized and difficult to transfer. Organizations, incentives, and business models evolved to manage those constraints.

With the advent of generative AI, something fundamental has shifted.

The signals are subtle at first: decisions moving closer to the edge, systems acting before humans notice, learning happening inside the process instead of after the fact. Taken together, they point to a different operating reality.

This is not simply a new technology wave. It is a change in the physics of business.

Leaders who sense this shift are not asking how to adopt AI. They are asking a more consequential question: *What assumptions no longer hold? What is suddenly possible that was not before?*

What was once available only to companies with billion-dollar budgets is now accessible to every business—large or small, digitally native or still analog.

Part I introduces this new physics, why it matters, and what it changes. Think of it as learning a new language before trying to write in it.

Chapter One

The Ground Has Shifted

The Smartphone we called a "Phone"

When the iPhone launched in 2007, we used it like a "phone," mostly for calls, email, and texting. We had a powerful computer in our pockets but were still thinking in terms of a monthly calling plan. It took years to realize its true power: it wasn't a better phone, but a camera, a map, a library, and a studio that could reinvent everything from social connection to global commerce.

The question wasn't, *"How do we make a call better?"* But *"What can we now build that was impossible before?"*

We are making the same error with AI.

The Era of Cognitive Scarcity is Over

Think about the last approval you waited for. That delay wasn't bureaucracy—it was scarcity. Someone with judgment had to read, analyze, and decide. And they could only handle so many requests per day.

For more than a century, businesses have been designed around one simple fact: **human cognition was the scarcest, most expensive resource in the system.**

Every process you've ever followed, every approval layer you've ever navigated, every quarterly plan you've debated was engineered around that one scarcity. We built hierarchies to distribute thinking. We created procedures to reduce mistakes. We batched decisions to conserve mental bandwidth. We accepted latency as a cost of doing business.

The entire architecture of the modern company—from the assembly line to the org chart to the annual budget—was built on the same constraint: *thinking was slow, expertise was rare, judgment didn't scale.* Scarcity.

That Constraint has Dissolved

AI removes the hard limit on the availability of human cognition—allowing us to infuse intelligent judgment at a scale and speed previously unimaginable.

The new scarce resource is no longer operational cognition—the ability to analyze, decide, act. It is strategic judgment—knowing which decisions matter and attending to them. AI doesn't simply accelerate our old systems—it replaces their underlying operating principles. The constraint is no longer how many decisions we can make, but **how many we are willing to let the system make for us.**

Most leaders today are using AI to do the wrong thing better. They are automating steps in legacy workflows. They are bolting chat interfaces onto rigid software. They are chasing efficiency gains inside a game whose rules have already been rewritten.

This is the small game.

The New Capability is Abundance

The big game begins when you stop trying to optimize the old world and start designing for the new one. This book is about that new **AI-native** world, built from the ground up around intelligence, not just using AI as a tool.

We are entering a world where:

- **Judgment can be distributed** *to where the data lives.*

- **Decisions are made continuously,** *not in meetings.*

- **Workflows adapt in real time,** *not during version updates.*

- **Systems learn** *with every interaction.*

- **The cost of trying something new** *approaches zero.*

But to harness it, you must first understand its nature.

Why "Probabilistic" is Human

For decades, enterprise software was **deterministic**: *IF transaction > $10K, THEN flag.*

Simple, rigid, and blind to nuance. Our entire operating architecture was built on this bedrock of predictable calculation. But AI systems are not deterministic; they are probabilistic.

Leaders often dismiss AI's probabilistic nature as "unreliable" for business. But that's how all systems designed around human cognition operate. AI joins this world of human-like cognition, not machine-like determinism.

Let's take some examples.

A doctor doesn't *guarantee* a diagnosis—they assess symptoms, weigh probabilities, and recommend a next step with high confidence. A pilot doesn't know *exactly* when turbulence will hit—they interpret radar, weather patterns, and experience to navigate safely. This is probabilistic reasoning.

AI does the same. Instead of a brittle rule, it delivers a **reasoned judgment**:

- **Instead of:** "Customer qualifies for loan: YES/NO"

• **AI provides:** "92% confidence this applicant will repay, based on 12 behavioral patterns. Recommended: approve with standard terms. Review key factors."

It's not introducing uncertainty; it's *quantifying* the judgment that your best experts already apply. The shift isn't from certainty to risk—it's from blind rules to **informed, auditable confidence**.

Where your business needs deterministic outputs, you can impose software-like rules: *same input → same output, every single time.* But where your business needs intelligence that can deliver better judgments faster, embrace the probabilistic layer as your new competitive cortex for decision-making. Build your governance around managing confidence, not chasing certainty, and you'll deploy intelligence that is both more powerful and more human.

This isn't a technology shift. It's a structural shift in capability. It is as fundamental as the move from manual labor to mechanization or from local commerce to global supply chains. Just as the internet dissolved the scarcity of distribution, AI is dissolving the scarcity of cognition.

The ground has shifted. You are now standing on a terrain of **abundant intelligence**.

This book is about how to **think and rebuild** your business—your products, your operations, your economics, your very strategy—on the new foundation where intelligence is abundant.

☆ ***Street Rule:*** *If intelligence becomes abundant, every business built on its scarcity must be redesigned.*

CEO Pitch: The Stakes Have Changed

"We must stop asking, *how do we add AI to what we do?* and start asking, *if intelligence were abundant and adaptive, what would we build from scratch?*

The distance between "experimenting with AI" and "being disrupted by it" is no longer decades, it's measured in product cycles, and the question isn't whether we'll lead the shift or be transformed by competitors who move before us."

Up Next

To understand what the new terrain of abundant intelligence will enable, we have to look through a lens of first principles. The answer lies not in one change but in nine interconnected shifts that are quietly rewriting the rules of business.

Chapter Two

The Nine Capability Shifts

The Store that Watches you Think

When Amazon launched its cashier-less Amazon Go stores, most observers focused on the convenience: grab what you want and walk out. But that wasn't the shift.

The invisible shift was this: the store itself became a **distributed sensing system** that could track intention, hesitation, comparison, and satisfaction in real time. The system didn't just eliminate checkout lines—it dissolved the boundary between browsing behavior and inventory intelligence, between customer preference and supply chain response.

The new capability wasn't efficiency—it was **continuous, contextual understanding at scale.** If you don't notice this shift, you'll make the same category error with AI: *optimizing for faster checkouts* while missing the complete *reinvention of retail intelligence.*

It is time to look at the nine invisible shifts that enable the new architecture of AI-native businesses in the age of intelligence.

The Capabilities of Abundance

What follows are nine fundamental shifts that provide a lens for **what's possible.** These shifts dissolve a constraint that shaped businesses for generations and open up the possibilities to reimagine your business. Let's examine each shift.

Shift 1: Scalable Reasoning—*AI applies judgment at a speed, scale, and consistency beyond human limits.*

We can multiply the decision-making ability of functional experts at machine speed across millions of transactions and events. This collapses decision latency and redefines operational intelligence.

Shift 2: Cheap Cognition—*The marginal cost of thinking (analysis, simulation) approaches zero.*

We can ask more questions, run more scenarios, simulate every anomaly, and correct course faster than ever. We can stop thinking in monthly, quarterly, yearly cycles and start thinking continuously.

Shift 3: Per-User Systems—*Products adapt to individuals, not averages defined by segments.*

We can stop thinking one-size-fits-all and understand people as individuals, not segments. We must shift from designing *products* to cultivating *adaptive experiences.*

Shift 4: Co-Evolving Agents—*Software stops being a tool and becomes a partner that learns with you.*

We are no longer deploying software but a learning system that improves with every interaction, correction, and edge case. Our advantage will compound as our system absorbs real-world experience.

Shift 5: Knowledge Without Expertise—*AI erases the competence bar to perform expert-level tasks.*

We can democratize expertise, elevating non-experts to perform expert-level work without being bottlenecked by scarce talent.

Shift 6: Intent Inference—*Systems understand what you're trying to accomplish, not just what you clicked.*

We can design for outcomes and let the system understand ambiguous instructions to infer real intent. User friction will plummet when systems think like humans, not the other way around.

Shift 7: Language as Interface—*Conversations replace* menus, forms, and buttons.

We can make our systems intuitively accessible to anyone, regardless of technical skill and without the need for onboarding. Conversations will replace menus and manuals to unlock faster adoption.

Shift 8: Probabilistic Trust—*When AI behaves probabilistically, trust must be engineered, not assumed.*

We must architect trust into intelligent systems from day one—with explainability, audit trails, and human oversight as foundational features, not afterthoughts.

Shift 9: Learning-Based Scale—*Scale is no longer about headcount or footprint but about learning velocity.*

We can outsmart giant competitors with small teams. Their massive budgets cannot replicate the compounding advantage of learning velocity.

How the 9 Shifts build on each other

The nine shifts aren't independent capabilities. They build on each other in a specific sequence to embed autonomous judgment into the fabric of your organization.

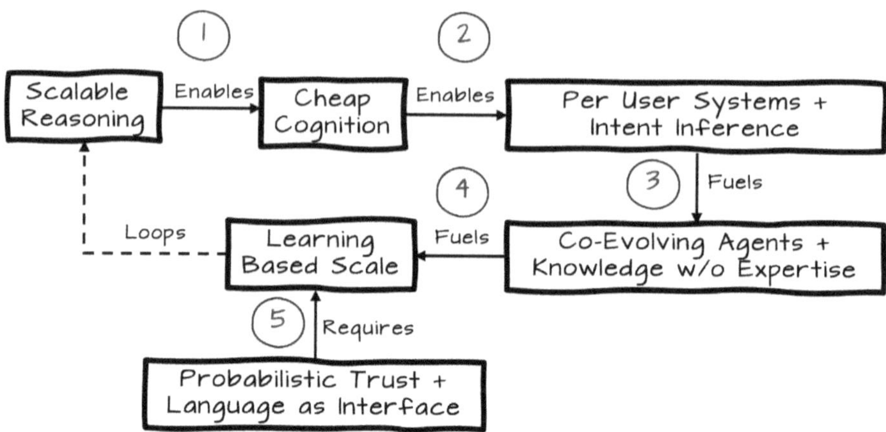

1. Scalable Reasoning makes Cheap Cognition possible. When AI can apply judgment across millions of transactions simultaneously, the marginal cost of each additional decision approaches zero.

2. Cheap Cognition enables Per-User Systems and Intent Inference. When thinking is cheap, you can customize every experience individually and interpret messy human intent at scale—creating a feedback loop of personalization and understanding.

3. Per-User Systems and Intent Inference generate data for Co-Evolving Agents and Knowledge Without Expertise. Every personalized interaction creates learning signals. This fuels systems that learn with you and democratizes expertise across your organization.

4. Co-Evolving Agents and Knowledge Without Expertise fuel Learning-Based Scale. As agents learn and expertise distributes, value compounds with usage: more users → more learning → better system → more users.

5. Learning-Based Scale demands Probabilistic Trust and requires Language as Interface. When systems operate probabilistically at scale, trust cannot be assumed—it must be architected. This complexity stays hidden behind natural conversation.

The nine shifts together unlock something entirely new: the foundation of AI-native systems that enables you to delegate decisions to autonomous systems safely. That's **Micro-Autonomy.**

Micro-autonomy is the smallest unit of work a system can perform independently—with context, judgment, and without human orchestration. It enables micro-decisions, which once required human judgment, to run autonomously at scale.

Micro-Autonomy: Why this matters?

There are two approaches to autonomy. **Macro-autonomy** delegates entire workflows, journeys, or outcomes as a single act of control. **Micro-autonomy** decomposes that same delegation into many small, localized decisions that are individually autonomous but cumulatively produce transferable judgment.

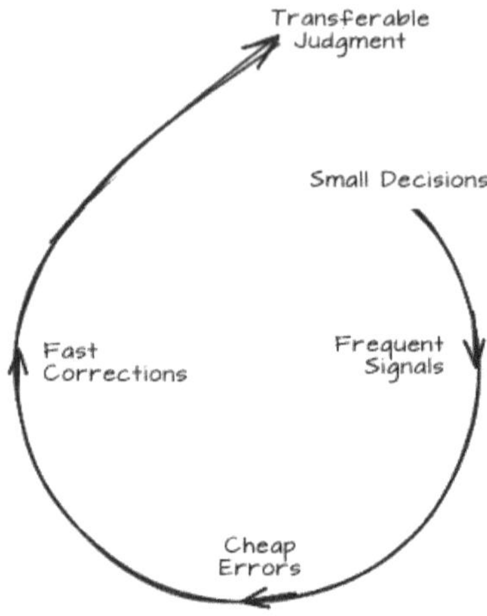

The difference is in one large bet versus thousands of small experiments—each cheap, informative, and reversible. Both aim for the same destination, but only micro-autonomy gets smarter on the way there.

Here's why: Macro-autonomy produces feedback only at completion. If a project takes 6 months, you get two learning cycles per year. Micro-autonomy generates feedback with each small decision—potentially thousands of cycles in the same period.

The improvement depends on feedback density—its frequency and resolution. And feedback density requires autonomy that is **granular**.

Case: Travel Expense Management

An organization sets a single autonomous travel policy: Departmental travel budgets, per-diem limits, pre-approved routes and vendors. The system's job is simple: *keep total spend within budget.* When costs spike, leadership only sees the outcome: *Travel exceeded budget by 18% this quarter.*

What they **don't** know: Was it last-minute bookings? Poor timing? Vendor price volatility? Or Certain routes or teams behaving differently?

The feedback is late, aggregated, and blunt. The response is equally blunt: tighter limits, more approvals, and frustrated employees. The policy is rule-based, it does not learn.

Now imagine micro-autonomy applied here. Instead of rules to "manage travel spend," the system makes **many small, local decisions** at the moment of action:

- Should this flight be booked now or delayed 24 hours?
- Is this hotel price normal for this city and date?
- Should the traveler be nudged toward a different route or carrier?
- Is this trip likely to generate enough business value to justify premium options?

Each decision:
- is contextual (role, urgency, client value, timing),
- produces immediate feedback (price delta, policy drift),
- is cheap to reverse, and is logged with outcome and confidence.

A single expensive trip doesn't "break the budget." It sharpens the system's judgment about *when* higher spend is justified.

Who should See these Shifts?

These shifts apply whether you are:
- a CxO of a large enterprise navigating transformation,
- a business unit leader trying to re-architect operations,
- a product leader designing the next generation of intelligent systems, or
- a founder building a new AI-native startup.

Now that you have seen these shifts working together, you cannot unsee them. If you absorb these nine shifts, the rest of the book will make the path obvious.

☆ ***Street Rule:*** *If you can't name which of these nine capabilities your AI pilot is harnessing, you're hanging a painting in a house that's on fire.*

CEO Pitch: The Mental Shift

"We have been measuring our AI efforts by cost savings and efficiency gains. That is like measuring Galileo's telescope by how much cheaper it made candlelight. We are not looking at better light. We are looking at a new solar system. Our strategy must be rewritten for this new universe of capability."

Up Next

The nine shifts are raw capabilities. But capabilities don't create business value until they're organized into a decision-making structure—one that makes autonomy safe, scalable, and self-improving. It's time to meet the architecture that turns these shifts into micro-autonomy that transforms your business.

Chapter Three

The Three Engines of Micro-Autonomy

Fewer, not Faster

The manager of a regional bank's loan department had a problem. Her team was drowning in applications. "We need to process these faster," she told her tech team. "Can AI help?"

The tech team did what they were asked. They built a brilliant system that read applications, extracted data, and populated the same old checklist 300% faster. The team was still drowning. They were just drowning in a faster, more efficient queue.

Automating steps, not rethinking decisions, is the trillion-dollar error. They had asked the wrong question. In the age of abundant intelligence, the goal is not to **process** decisions faster. The goal is to **make fewer of them**.

The right question wasn't "How do we speed this up?" It was: **"What is the smallest, safest decision we can stop making altogether?"**

The manager asked for **automation**. She got a faster queue. The right question leads to **AI-native transformation**: redesigning the system so fewer decisions need to be made at all.

The Fork in the Road

Automation thinking and AI-native thinking are not different degrees of the same thing. They are different destinations.This chapter is not about automation. It is about AI-native architecture required for transformation. It begins with introducing the concept of **Micro-Autonomy**.

Think Micro-Autonomy

Look at any workflow in your company—customer onboarding, inventory replenishment, IT troubleshooting. It is a chain of human decisions: *Should we approve this? Is this the right part? Is this a priority one ticket?*

Our instinct is to build a system that helps a human make each of those decisions slightly faster. This is a dead end. It leaves intact the fundamental constraint, which is the bottleneck of human decisions in the workflows.

The fundamental unit of work in an AI-native business is not a task, a transaction, or a meeting. It is the delegated micro-autonomous decision.

It is not big, brittle "automation." It is not a chatbot. It is the **atomic particle** of intelligent action:

• A credit card system **autonomously flagging** a transaction as fraud.

• A support bot **autonomously issuing** a refund under $50.

• A logistics system **autonomously rerouting** a truck around a storm.

• A marketing engine **autonomously pausing** a low-performing ad.

Think of it as the moment you delegate a sliver of judgment, with clear guardrails, and the system executes it. Not with a rule ("if charge > $500, flag"), but with reasoning ("given this customer's pattern, location, and purchase history, this charge has a 92% probability of fraud—flag for review").

This is the center of gravity. Everything in this book—every system, every team structure, every economic model—exists to inject safe micro-autonomy into your operations.

The Three Laws of Micro-Autonomy

For a decision to qualify as micro-autonomous, it must satisfy these three laws:

1. It must exercise Judgment: Use reasoning, not just rules.
Your business should be able to analyze, simulate, and make decisions at scale and speed.

2. It must operate within Governance: Humans set clear boundaries and oversee exceptions.
Your business should be able to interact with people and businesses with trust and transparency.

3. It must Compound intelligence: Learn and get smarter with every execution.
Your business should get smarter with every interaction to create competitive advantages.

These three laws point to the **three engines that make micro-autonomy possible.** These three engines are the capabilities of the nine shifts organized to work together to create micro-autonomy. Think of the nine shifts as the foundational capabilities and the three engines as the systems you build using them.

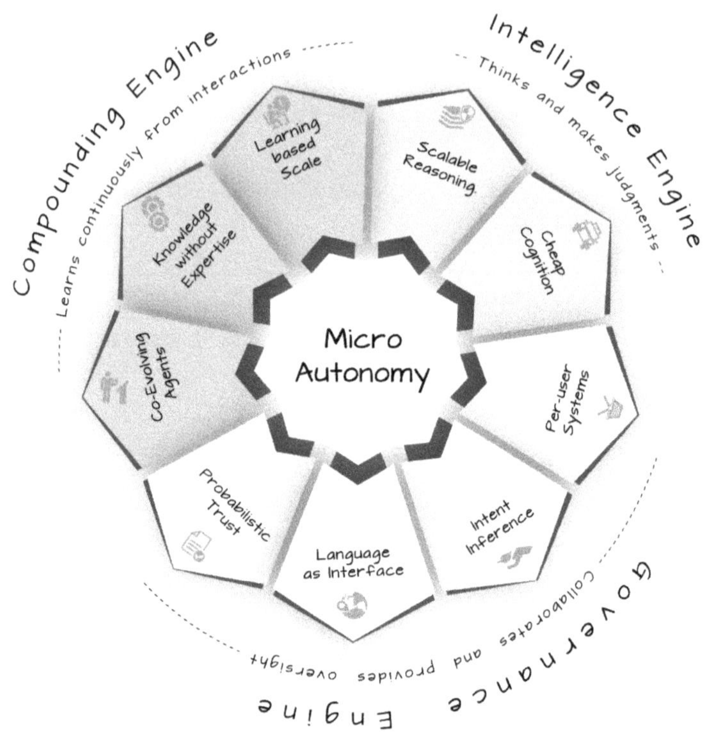

Three Engines of Micro-Autonomy

Let's see how each engine works, using daily life examples.

The Intelligence Engine

Example: Your Credit Card's Fraud Detection

When your credit card company spots suspicious activity, it's the Intelligence Engine at work.

• **Scalable Reasoning:** It analyzes millions of transactions simultaneously, comparing your spending patterns against global fraud trends in milliseconds.

• **Cheap Cognition:** Running these checks costs the bank pennies per customer—impossible with human analysts.

• **The Result:** You get a text: "Did you buy $249 at ElectronicsRUs in Tokyo?" within seconds of the transaction.

Now imagine that level of intelligence applied to:

• Your **supply chain** *predicting stockouts before they happen.*

• Your **marketing spend** *identifying underperforming campaigns in real-time.*

• Your **quality control** *detecting defects invisible to human inspectors.*

What it does: The Intelligence Engine analyzes signals, generates contextual recommendations with confidence scores, and explains its reasoning—all at a scale and speed humans can't match. It turns your business into a thinking organism.

The Governance Engine

Example: Your Bank's Loan Approval System

When you apply for a car loan online and get instant approval, it is the Governance Engine at work behind the scenes.

The Intelligence Engine analyzes your application and recommends: *"Good to approve $30,000 at 4.2%. Confidence: 88%."* Here's how the Governance Engine orchestrates that recommendation:

• **Per-User Systems:** The system recognizes you're a first-time borrower (not a seasoned customer) and adapts its communication style—more detailed explanations, more conservative guardrails.

• **Intent Inference:** The system infers you're trying to buy a specific car and adjusts the offer to $25K to keep you within safe debt limits while still meeting your goal.

• **Probabilistic Trust:** The Intelligence Engine's 88% confidence is high—but Governance doesn't rely on that alone. It validates against **deterministic rules** (debt ratio limit), catches the violation, and escalates to human review despite the strong confidence score.

• **Language as Interface:** Instead of "Application denied: DTI ratio exceeds threshold," you get: "We can approve $25,000 at 4.2%, but your current debt-to-income ratio is 44%, which exceeds our 40% limit for autonomous approval. A loan officer will review your application within 2 hours to explore options."

The Result: The system blocks autonomous approval, escalates to a loan officer with clear reasoning, and maintains trust through transparent explanation. The loan officer reviews, determines you can qualify with a co-signer, and approves within hours.

Now imagine that applied to:

• Your **customer support** *personalizing responses, inferring true issues, explaining decisions clearly.*

• Your **order fulfillment** *adapting shipping options, catching unusual patterns, confirming before executing.*

• Your **medical scheduling** *assessing urgency per patient, inferring preferences, explaining wait times transparently.*

What it does: The Governance Engine personalizes interactions, understands true intent, routes actions based on confidence and risk, validates every decision against business rules, and communicates in natural language—making micro-autonomy safe and trustworthy. The Governance Engine provides oversight and manages uncertainty transparently **making AI safe**.

The Compounding Engine

Example: Your Email Spam Filter

Remember when spam filters were terrible? They'd block important emails and let junk through. Today's spam filters are incredibly accurate. And they get better **because you use them**:

• **Co-Evolving Agent:** When you mark something as "Not Spam," it learns. When millions of people do this, the filter evolves globally.

• **Knowledge Without Expertise:** You don't need to be a spam expert—the system has learned from billions of emails what "spam" looks like.

• **Learning-Based Scale:** Every user makes the spam filter smarter for all other users. *More users = more corrections = better filtering.*

Now imagine that level of continuous improvement applied to:

• Your **sales process** *learning which approaches close deals, improving with every interaction.*

• Your **product recommendations** *getting better at predicting what customers want.*

• Your **operational workflows** *continuously optimizing routing, scheduling, resource allocation.*

What it does: The Compounding Engine captures every decision outcome, learns from what worked and what didn't, and continuously improves the Intelligence Engine—creating advantages that compound over time and can't be replicated with capital alone.

Hallucination: The Elephant in the Room

This is where the Governance Engine earns its keep. The Intelligence Engine's great strength is generative reasoning, but this comes with an intrinsic trait: statistical uncertainty. It can hallucinate, guess, or pattern-match its way to a plausible-but-wrong answer. This isn't a bug, but that's how statistical learning works.

The Governance Engine's architecture is the antidote. It operates on a simple principle: *Creativity in reasoning, correctness in action.* It allows the Intelligence Engine to explore possibilities but imposes a series of validations before any decision becomes real:

• **Guardrails**: Outputs are constrained to valid ranges.

• **Tiered autonomy**: Confidence → Acts, Suggests, Asks.

• **Human oversight**: Low-confidence decisions escalate.

• **Explainability**: Decisions can be traced and validated.

The goal isn't to eliminate LLM creativity in the Intelligence Engine, that's the source of their power. The goal is to orchestrate generative flexibility toward correctness before they become actions.

This is why governance is non-negotiable. Micro-autonomy without it is chaos. Intelligence without oversight is dangerous.

Data is Substrate, not Intelligence

In AI-native systems, data doesn't "drive decisions" but conditions judgment.

Enterprise systems (ERP, CRM, Data Lakes) provide facts; engines provide reasoning.

In the diagram below, the Judgment Layer is the architecture, the three engines are the mechanism, and micro-autonomy is the behavior they produce.

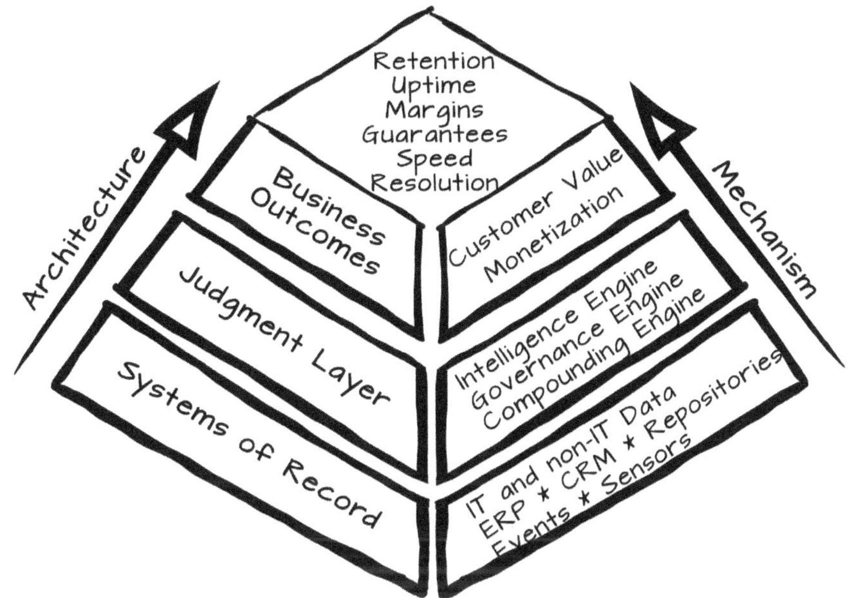

Intelligence Engine: Uses data to reason and propose actions
Example: Customer history | order status → recommends next best action.

Governance Engine: References data to authorize, limit, or escalate
Example: Contract terms + account tier → validates or blocks action.

Compounding Engine: Learns from outcomes to improve judgment
Example: Which decisions resolved issues → recalibrates confidence thresholds.

Data provides facts. Engines provide judgment. Outcomes create value.

Intent: The Human Compass

You have the engines. You have the data. But who tells it what to do?

As systems take on judgment, **intent** defines what success means: the objectives, priorities, and constraints set by people. Without clear intent, autonomy is indistinguishable from automation. Let's see how intent operates across the three engines.

Intelligence Engine → Uses intent to reason
Example: "Resolve issue while preserving lifetime value" → evaluates refund/replacement/escalation.

Governance Engine → Enforces intent in action
Example: "High-value customers get flexibility" → allows refund for premium, blocks for risky.

Compounding Engine → Learns from intent outcomes
Example: Observes which actions meet intent → adjusts confidence thresholds.

The Three Engines Flywheel

When the three engines converge to enable micro-autonomy, they create a flywheel effect.

Intelligence enables small autonomous decisions → Governance bounds them safely with trust → Compounding learns from outcomes → Intelligence adapts, uncertainty reduces → Micro-autonomy expands and sharpens.

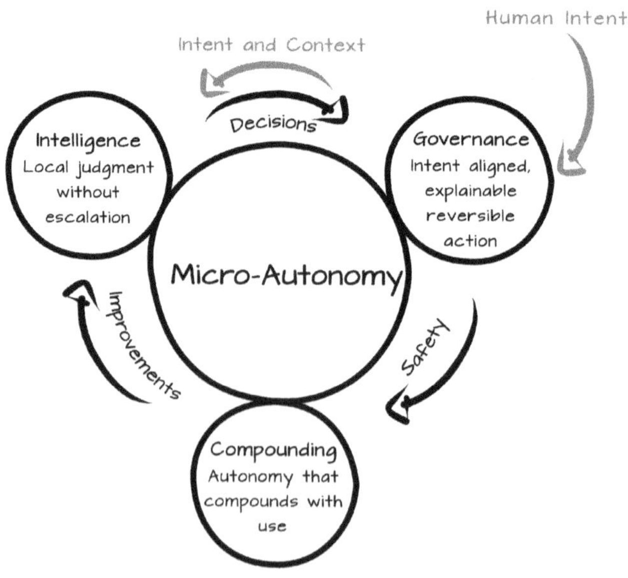

This flywheel shows how micro-autonomy compounds over time. Each turn of the flywheel increases the quality, safety, and speed of autonomous action—allowing the organization to scale decision-making without centralizing control or slowing execution.

Here's the critical insight: **Micro-autonomy only happens when all three engines work together.** Remove any one, and you don't have micro-autonomy but an incomplete automation:

• **Intelligence without Governance** = System no one trusts *(too opaque, no override, alienating)*.

• **Intelligence without Compounding** = Fixed system *(static accuracy, no learning from mistakes)*.

• **Governance without Intelligence** = Fancy interface but no substance *(looks good, but dumb)*.

• **Compounding without Governance** = System without oversight *(drifts unpredictably, no accountability)*.

Micro-autonomy in Common Workflows

• **Sales:** *Should we follow up with this lead?*
Intelligence: analyze engagement patterns; Governance: confidence threshold for auto-outreach; Compounding: track response rates.

• **Operations:** *Should we reorder this part?*
Intelligence: analyze usage + lead time; Governance: set auto-reorder thresholds; Compounding: learn from stockouts.

• **Support:** *Should we escalate this ticket?*
Intelligence: analyze complexity + urgency; Governance: define escalation criteria; Compounding: learn from resolution outcomes.

Case: Customer Support

Let's see all three engines working together in a real workflow of customer support. Each incoming customer interaction becomes a candidate for micro-autonomy: draft a response, adjust tone, issue a refund, offer a workaround, escalate immediately, or do nothing.

1. Start with the Intelligence Engine: Micro-autonomy reasons through context—purchase history, contracts, usage patterns, sentiment signals—this makes autonomy *situational.*

Interprets message context. Proposes micro-actions: draft response, adjust tone, issue refund, offer workaround, recommend escalation, OR do nothing. *What is the best action here, and how confident am I?*

2. Activate the Governance Engine: Every interaction is personalized and in natural language. The AI infers intent and adapts its approach and uses confidence thresholds for human escalation.

Personalizes by user or risk level. Ensures explainability. Decides on action: refund requires approval, escalate to human, sensitive account mode. *Given this customer, risk, and confidence—may this action be taken?*

3. Engage the Compounding Engine: Each support outcome—resolved, escalated, or reopened—feeds back into the system to get smarter with every interaction.

Logs outcomes: resolved, reopened, escalated. Tracks customer satisfaction. Learns: actions worked in which contexts. Improves future action selection. *"What did we learn, and how should future decisions improve?"*

You just turned a support workflow into a self-tuning decision network. Costs fall. Satisfaction rises. And the system learns how to act better tomorrow than it did today.

Getting Value out of Micro-Autonomy

When designing for micro-autonomy, ask yourself:

• Are we automating a known workflow or creating a process that was previously impossible?

• Are we improving digital workflow or giving the analog native their first intelligent system?

• Are we building it for existing users or to turn non-participants into empowered actors?

• Are we building to empower domains that have operational data but lack intelligence to harness it?

This distinction separates incremental improvement from strategic transformations.

The Maturity: From Automation to Autonomous

Not all work is ready for micro-autonomy. It exists on a spectrum, from tools to true partners. The maturity scale below is about progressively transferring judgment to systems where confidence, governance, and learning make autonomy safe.

• **Level 1: Assisted Tools**—Human decides, System assists
Example: Spell-check suggests; you choose.

• **Level 2: Rule-Based Automation**—System acts on static rules
Example: Email filters based on sender.

***The Chasm** - This is where the AI opportunity begins. It's the shift from "if this, then that" to "given this situation, the best action is..." This is the leap from Automation to AI-native Transformation—from following rules to exercising judgment using micro-autonomy.*

- **Level 3: Contextual Suggestions**—AI recommends, human decides
"High fraud risk—recommend review" with explanation.

- **Level 4: Supervised Autonomy**—AI acts when confidence is high (this is the initial **Target State)**
Fraud system auto-blocks obvious fraud.

- **Level 5: Self-Evolving Autonomy**—AI acts, decides, learns, improves
Recommendations refined based on user behavior.

- **Level 6: Strategic Autonomy**—AI proposes strategy; humans choose direction
"Enter new segment X based on pattern analysis."

Most companies today operate at Level 1-2 and few with Level 3 experiments. The transformation begins with Level 4-5 where 80-90% of micro-decisions happen autonomously, experts focus on exceptions and learning loops compound.

> ☆ ***Street Rule:*** *If your AI project doesn't delegate a single new decision to the machine, you're just doing expensive analytics.*

CEO Pitch: The New Competitive Vector

"We've competed on operational efficiency for decades. That game is ending. The new game is **delegated judgment**: how much intelligent decision-making we can safely transfer to the system, and how quickly that system gets smarter with every delegation. While our competitors automate old processes, we must transform our decision architecture. We're playing a different game entirely."

Up Next

How do you reimagine your business model around the three engines? This is where the new operating system turns your traditional business into an intelligent organism.

Chapter Four

AI-Enabled Business Autonomy

The Guarantee that Broke the Market

Imagine your company sells enterprise CRM software. Your business model has worked for 20 years: per-seat licensing, implementation services, annual maintenance. You're profitable, with high CSAT and NPS scores.

Then you start losing deals that you thought were locked in. A competitor one-tenth your size is winning with an offer that makes no sense to your finance team: *We guarantee a 15% reduction in your customer service costs. If we don't deliver, you pay nothing.*

Your sales team calls it a desperate gimmick. Your finance team says the unit economics are impossible. Your product team says guaranteeing outcomes for every customer is mathematically unfeasible—every customer is different.

Yet the customers who've been with you for a decade are switching. You have no evidence their product is any different. The business model that made you successful is suddenly making you vulnerable to a competitor that you wish did not exist.

What just happened? Let's find out.

Your Business is being Reimagined.

Here's what didn't happen: Your competitor didn't build a fundamentally different product. They didn't invent a new category of enterprise software. They didn't replace the traditional business model.

Here's what did happen: They reimagined their business around **autonomy** embedding intelligence in the same product category you're in. They turned their software from a tool into a platform that delivers **guaranteed outcomes**.

The software itself? Still enterprise software. Still has features, interfaces, workflows. But now it's the **delivery vehicle** for something more valuable: **autonomous intelligence that guarantees results.**

This is the shift leaders should not miss. AI doesn't destroy your business model. It doesn't require you to scrap your products and start over. *It embeds autonomy that transforms what your existing products and services can deliver.*

Think of it like this:

• **Before:** You sold a product (software, machine, service).

• **After:** You sell the same product, but with embedded autonomous intelligence that delivers **outcomes.**

The offering becomes the **platform** and the business model **evolves** to capture new value.

Business Model and Autonomy

Business model is the *underlying logic* of how a company survives and thrives. It is the system that converts insight into value, value into revenue, and revenue into reinvestment and advantage. It is built on five foundational interconnected blocks:

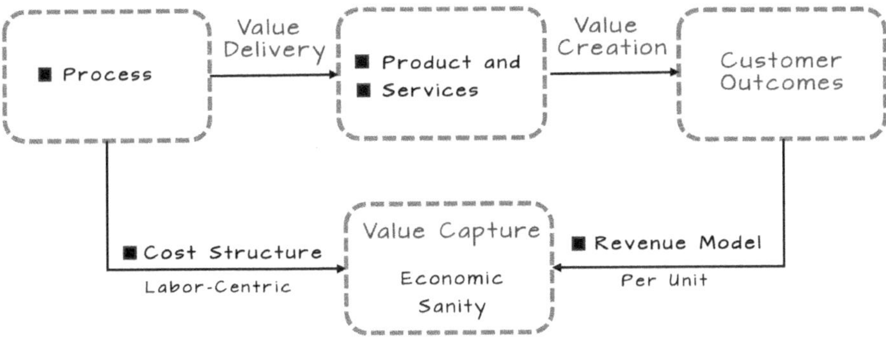

1. Product (The Artifact) *What you sell.* A physical good, a software license, a defined service bundle.

2. Services (The Human Layer) *How you help people use it.* Support, training, implementation, customization.

3. Process (The Recipe) *The sequence of steps to deliver value.* From procurement to assembly lines to distribution.

4. Cost Structure (The Spend) *Where the money goes.* Labor, materials, assets, utilities, compliance, overheads.

5. Revenue Model (The Earn) *How you get paid.* Per unit, per seat, per hour, per subscription.

For a century, this architecture was stable. Departments optimized their own blocks: finance managed cost, sales managed revenue, and operations managed process. The model was a **machine**—powerful, predictable, and rigid.

This model still works. But the 9 shifts of AI acting through three engines enable AI-native micro-autonomy that changes everything. It is not a bolt-on feature. It is a capability that lets you reimagine your business model around intelligence—embedding it in your products, services, and processes. The possibilities are now unbounded to build entirely new business models.

Let's see how.

How AI-native Autonomy transforms Products

For decades, a "product" meant a static artifact with a bundle of features for a target segment updated through version releases, designed around UX and structured around human steps. The AI-native micro-autonomy changes the equation.

Product is no longer the code or the hardware. It is now a delivery vehicle for outcomes enabled by the intelligence of micro-autonomy.

Five Dimensions of an AI-native Product

1. Product as Intelligence: The product's core value is the **reasoning** embedded within it. A financial tool that reasons about portfolio risk. A healthcare system that reasons about patient signals.

2. Product as Individualized Experience: The system adapts to every user, every context, every moment. Your product becomes "one product per user." It delivers customer specific outcomes. This is the most underrated economic shift of AI that can bring individualized monetization.

3. Product as Autonomous Actor: AI can now *act*—detecting, predicting, initiating, escalating, even fixing. A fundamental redesign of product capability.

4. Product as Learning Partner: Product is no longer static, it is a learning organism that compiles experience, exhibits behavior. User interaction improves the system, correction sharpens behavior, novices perform like experts.

5. Product as Conversation: Conversation is a new interaction layer that takes you from intent to interpretation, concerns to explainability.

Example: A CRM doesn't just store contacts—it predicts which deals will close, suggests optimal outreach timing, and gets smarter with every sales interaction. The product *is* the prediction engine.

The shift: Your product becomes the **platform** that delivers autonomous intelligence to every customer.

How AI-native Autonomy transforms Service

In most companies, service is still defined as answering tickets, deploying products, solving problems once they've already happened. It needs humans to interpret the nuance. That era is over.

The micro-autonomy makes service continuous and embedded. It moves from **reaction** → **prevention** → **orchestration**. And once leaders understand this shift, the entire business model begins to tilt.

Every touchpoint (a tap, vibration, anomaly, pause, delay, pattern) feeds an insight into what the customer needs next. A tractor knows soil conditions before the farmer asks. A turbine knows the likelihood of failure weeks in advance. A hotel room adjusts climate before a guest touches a switch.

Service no longer waits for a trouble ticket. It detects patterns before humans notice symptoms and can apply fixes autonomously, invisibly, and continuously. And once service becomes anticipatory, its relationship to the customer changes from "fix my problem" to "protect my outcome."

This is a major reframe: *Service stops being the ambulance, it becomes the immune system.* The real transformation begins when service is no longer a function, but a capability embedded in the product itself.

Example: A bank's "customer service" isn't a call center—it's an AI that monitors your account 24/7, detects unusual patterns, proactively suggests actions, and explains every recommendation in plain language. Human bankers handle exceptions and complex judgment calls.

Every service interaction expands the system's understanding. It sits at the intersection of product usage, customer behavior, environmental context, operational constraints, emergent issues, and real-world edge cases. It becomes the training ground for intelligence in the product.

This is why in AI-native business models service is not a cost but a compounding advantage. When service attains higher levels of assurance, you're no longer selling support—you're selling certainty. The service dictates the business model—think uptime guarantees, productivity-based pricing.

The shift: Service becomes **continuous, embedded intelligence** that the product delivers.

How AI-native Autonomy transforms Process

For most of business history, a process was a map to represent workflow. Maps and workflows existed because humans needed instructions and teams needed coordination. These maps were guardrails to check against human errors.

Companies were organized around who did what, when, and why. Pathways in healthcare, routing flows in logistics, SOPs in manufacturing, replenishment cycles in retail.

Micro-autonomy renders this logic obsolete. Processes engineered for human pacing and error reduction are replaced by intelligent, contextual flow. Why have a 14-step loan approval process when AI can evaluate risk in milliseconds? Process dissolves into self-optimizing flow.

Traditional processes are linear: *if A happens, do B, if B fails, escalate to C.*

AI-native processes are **nonlinear**: they interpret context, infer intent, evaluate thousands of micro-signals, and decide the next optimal action; often one the original workflow designer never envisioned.

A construction inspection process becomes a computer-vision engine that detects structural deviations, safety hazards, and compliance issues before the site manager even walks the floor. A warehouse workflow becomes real-time optimization that reroutes pallets based on congestion, shelf conditions, staffing levels, and forecasted shipments.

Processes built for humans carry friction by design. Intelligence operates at a different tempo. The real breakthrough is not automating tasks—it is **removing the tasks altogether**.

In the AI-native world a process is no longer a sequence of tasks. It is a network of human and machine agents collaborating through shared intelligence. It continuously senses, interprets predicts, adapts, integrates. And gradually processes begin to **self-heal**, **self-correct**, and **self-optimize**—properties no workflow ever had.

Example: In manufacturing, real-time computer vision models detect defects and adjust machine parameters autonomously. The QC process is thus absorbed into the production process itself. No handoffs—self correct, self-heal.

The shift: Processes become **self-optimizing systems** where intelligence is embedded.

Note: Process and Service areas are the best places to start with micro-autonomy experiments. Process loops offer repetitive workflows with clear inputs/outputs (invoice approval, loan underwriting, inventory replenishment), and Service loops offer customer-facing interactions with known patterns (inquiry routing, support tickets, scheduling, onboarding). They're repetitive, data-rich, and measurable—ideal proving grounds for the micro-autonomy experimentation.

How AI-native Autonomy transforms Costs

Cost has always been treated as allocations, depreciation, and unit economics. It is an accounting exercise to optimize it, control it, and even battle it during downturns.

Not all costs are expenses, many are penalties paid for the limits of human cognition. Labor exists because only humans could decide, buffers exist because forecasting was weak, inspections exist because machines couldn't interpret context.

With the micro-autonomy, the cost structure inverts. It doesn't cut costs—it **erases the constraints that create them.** The economics change from *labor-centric* to *intelligence-centric.*

By introducing real-time vision inspection, predictive maintenance, dynamic scheduling, and autonomous calibration—scrap plummets, downtime collapses, defects self-correct, energy adjusts to load, and labor shifts from repetitive work to supervision. Entire categories of cost collapse.

The Levers of Cost transformation:

• **Predictability** removes buffer costs by eliminating uncertainty.

• **Compression** shrinks cycle times autonomously—fewer cycles, fewer steps, fewer costs.

• **Reallocation** moves humans from low-leverage repetitive tasks to high-impact supervision and innovation.

Example: A $2M customer service operation (80 agents @ $25K loaded cost) shifts to $1.2Ma 40% reduction (30 agents supervising AI). Throughput rises 60%, (more inquiries resolved), cost per inquiry drops from $40 to $12, while customer satisfaction rises.

Start asking:

▸ What costs exist only because of delayed signals?
▸ What buffers exist because our predictions were too weak?
▸ What will our cost structure be if the entire operation can self-adapt?

Bottom line: Turn the cost structure into a competitive weapon.

The shift: Costs become a compounding advantage that decreases with scale as intelligence improves.

How AI-native Autonomy transforms Revenue

Revenue has always obeyed the architecture of the product. If you sold a machine, you charged per unit. If you sold software, you charged per seat. If you sold a service, you charged per hour. If you sold access, you charged a subscription. They lived in a locked marriage. You built the product, and the pricing followed its physical or functional boundaries.

Micro-autonomy breaks that link. The moment your product becomes intelligent, the economics stop following the product and begin following the **outcome**.

A predictive power grid system isn't valuable because of the meters but because it prevents outages. A medical imaging platform isn't valuable because it stores scans but because it elevates diagnostic accuracy. A logistics platform isn't valuable because it tracks trucks but because it reduces detention, delays, and unpredictability.

Once the outcome becomes the primary value driver, revenue must follow it and not the interface.

Three forces of that shape Monetization:

• **Intelligence deepens with time,** *revenue compounds with usage.*

• **Value decoupled from usage,** *tied to impact, not activity.*

• **Network leverage,** *every new customer improves outcomes for all.*

Together, these forces create new possibilities to monetize along new vectors of outcome: uptime, precision, safety, comfort, throughput, savings, productivity, compliance, stability, etc.

Example: A useful reference point is Palantir Technologies—a data analytics and decision-making company that works on high-impact decision domains from national security to supply chains, manufacturing, and financial risk.

Palantir Technologies built a $2B+ business on outcome-based ROI. They don't sell software licenses or consulting hours. They take projects where the business impact is quantifiable: reduce inventory turnaround from 23 days to 16 days, prevent $50M in fraud losses, optimize supply chain to eliminate $100M in delays. The ROI at project end justifies their cost.

This revenue model is enabled by micro-autonomy: *from selling tools to selling measurable outcomes.*

Start asking:

- What outcome does our product guarantee?
- Where does it provide the highest measurable value?
- How do we eliminate customer's uncertainty?

These questions produce entirely new revenue layers.

The shift: Revenue becomes **outcome-aligned.** You capture a share of the value the micro-autonomy creates.

Not Renovation, but a New Foundation

Most companies respond to AI by trying to add *AI features*. Micro-autonomy isn't a feature but a *capability* that transforms what your business model can deliver.

• You don't replace your products.
You embed intelligence that makes them platforms.

• You don't eliminate service.
You make it continuous and embedded.

• You don't remove processes.
You make them self-optimizing.

• You don't just cut costs.
You erase the constraints that create them.

• You don't abandon unit pricing.
You add outcome-based layers.

The traditional model was a blueprint for a world of **scarcity**:

Thinking was expensive → Batch it, ration it, centralize it
Expertise was rare → Hire specialists, protect them, scale carefully
Change was slow → Plan quarterly, execute annually, update in versions

The new model sells assured outcomes, delivered through ever-expanding micro-autonomy.

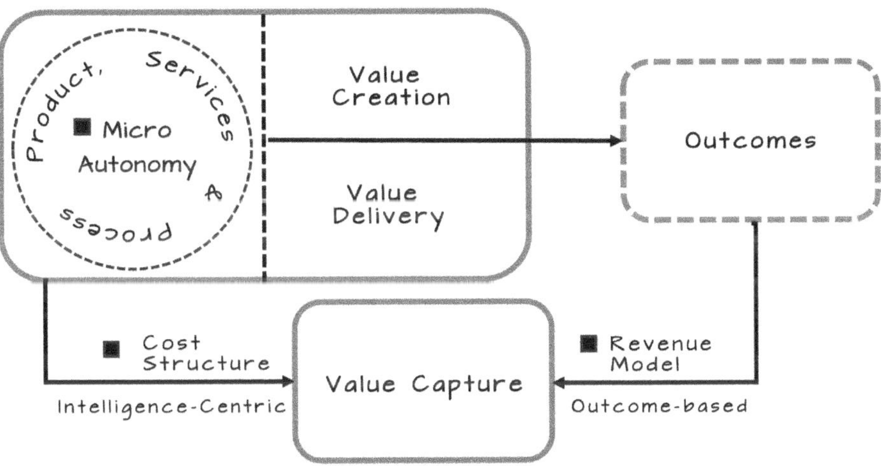

We now live in a world of **abundance**:

▸ Thinking is cheap → *Distribute it, apply it continuously, and let it compound*
▸ Expertise scales → *Embed it in systems, multiply it through AI, democratize it*
▸ Change is constant → *Sense continuously, adapt in real-time, evolve perpetually*

Your business model doesn't need to be rebuilt from scratch. It needs to be reimagined around autonomous intelligence.

A Note for Founders and Builders: *If you are building a company from scratch, the implications of this new model are even more radical. You have the unique advantage of starting with a blank canvas—no legacy systems, no organizational inertia. You can design your micro-autonomy in harmony from day one. For a detailed set of tactical principles on how to build from zero, see **Appendix A:** The Startup Builder's Playbook—The 7 Principles of AI-native Building.*

☆ ***Street Rule****: Your new business model isn't in a spreadsheet. It's in the answer to this question: "What autonomous capability could we embed in our existing products that would make outcomes—not features—the source of value?"*

CEO Pitch: Intelligent Autonomy for Business

"Our business model is a clockwork machine in a quantum world. The question is how to reimagine what we have around intelligent capability that transforms products into platforms, services into continuous partners, and revenue into outcome-sharing. The companies that will dominate the next decade will be built on the new physics of micro-autonomy that enables intelligence required to transform the existing business model."

The Transition

You've seen what the micro-autonomy does to each pillar of your business model. It doesn't destroy them—it transforms what they can deliver.

But adding an micro-autonomy capability isn't just a technology upgrade. It requires an entirely new architecture: one that generates intelligent decisions, governs them safely, and evolves continuously. That architecture is what Part II reveals.

Part-II: THE NEW FOUNDATIONS

The architect Louis Sullivan had a famous principle: *form follows function*. If the function changes, the form must change with it.

For a century, the function of a company was to execute repeatable processes efficiently. So we built organizations, business models, and economics around that function. Hierarchies for command-and-control. Products as artifacts. Revenue as units sold times price.

But when the function becomes *continuous learning at scale*, everything must be redesigned from first principles. Whether an existing digital business, a physical product company, or a digital revolution missed sector, the principle remains: you're not bolting intelligence onto a machine. You're reimagining a new game. Now comes the hard part: building the new architecture from first principles.

Part II shows you how to construct an AI-native business as an intelligence organism. One that doesn't just use AI but is fundamentally rebuilt around it. And why half-measures guarantee defeat.

Chapter Five

Business As An Intelligence Organism

The Vending Machine vs. The Garden

Until now we built businesses like vending machines. You put in a request (money, order), a predefined internal process clunks into action, and out drops a product. Consistent. Predictable. Over. Dead.

The AI-native business is a **garden.** You plant seeds (customer intent, data), and a living system tends to them—watering here, pruning there, adapting to weather, bearing fruit that is unique to each plant. It is dynamic, adaptive, and alive.

The vending machine has a mechanistic blueprint: levers, chutes, and coils. The garden has a biological blueprint: sensing, responding, and growing systems. We are shifting from the first to the second. You upgrade the vending machine by installing the new capability of micro-autonomy that makes it sense, reason, decide, and learn.

Now let's see how the pillars of the traditional business model transformed by micro-autonomy recombine into a higher-order architecture—one that doesn't just execute but evolves.

Your new blueprint is for an intelligent organism.

The AI-native Blueprint

We have seen how the foundational blocks (Product, Service, Process) of the business model transform in the AI-native world. Let's see how they recombine at a higher level as a single organism.

Value Creation = custom reasoning + learning-driven experience
(What you sell and how it gets smarter)

Value Delivery = personalized, trusted outputs via natural interface
(How customers interact with your offering)

These two layers correspond directly to the three engines we explored earlier.

The blocks of **Cost** and **Revenue**, as always, define economics, but differently.

Let's call it the Economics Engine.

Value Capture = micro-autonomy cost, outcome-aligned revenue.
(How you thrive, sustain and grow)

Here the distinction is about how the business model functions when enabled by the three engines of micro-autonomy. Let's see that.

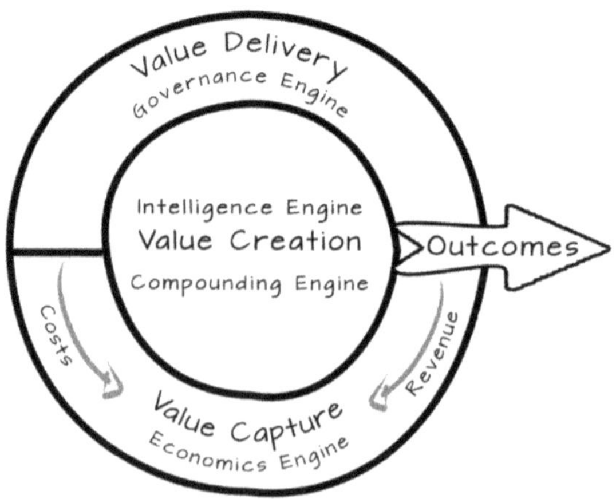

1. Value Creation (Intelligence+Compounding Engine)

In the AI-native business model, Product and Process fuse into the Value Creation Engine, where embedded intelligence and learning processes generate value together.

This is **what you sell and how you create value** with micro-autonomy and adaptive outcomes instead of static artifacts. It's a system that autonomously achieves your customer's outcome.

New paradigm: You're not selling tools. You're selling intelligent outcomes that compound.

Case: Health Platform

The Product and Process: Wearable devices + monitoring app (the hardware and software).

Intelligence Engine: The product that continuously analyzes health signals, detects early warning signs, predicts outcomes, and generates proactive intervention plans.

Compounding Engine: The process of continuous feedback loop that makes the platform smarter as it learns from every patient outcome and intervention result. If a predicted heart issue didn't materialize, it refines its threshold. If a recommended lifestyle change led to measurable improvement, it reinforces that pattern.

What you're selling: Not the wearable. Guaranteed health outcomes through a system that gets better at preventing illness. The device is the delivery vehicle for autonomous, learning health management.

2. Value Delivery (Governance Engine)

In the AI-native business model, the human-facing functions of Service recombine into the Value Delivery, becoming the very interface through which trusted outcome is collaboratively delivered.

It delivers an experience of safe actions that adapts to the customer's intent through a conversational layer. It builds the trust required for autonomous action through transparency and control. It's where Human-in-the-Loop participates in the Value Delivery.

New paradigm: Service is no longer a separate human layer. It is embedded intelligence that collaborates with each individual.

How it applies to Health Platform Case

The Service: Nurse hotlines, doctor consultations, care coordination.

Governance Engine: The trusted interface that collaborates with each patient. It validates every decision from the Intelligence Engine, explains reasoning in natural language, and routes actions appropriately based on confidence and context.

How it governs:

• For high-confidence decisions (>90%): ACT autonomously.
Example: "Your heart rate pattern is unusual. I've scheduled a same-day cardiology appointment. Here's why..." (Patient can override.)

• For medium-confidence decisions (70-90%): SUGGEST and await approval.
Example: "I notice a weight gain trend. Consider adjusting protein intake. Would you like me to generate a meal plan?"

• For low-confidence decisions (<70%) or rule violations: ESCALATE to a human.
Example: "Multiple concerning signals detected for this patient. Here's what I've observed..."

What you're delivering: Not reactive care, but proactive, trusted partnership that embeds care into daily life. The human medical team is still there but now focused on complex judgment, empathy, and exceptions, while the Governance Engine handles routine monitoring, early detection, and coordination with transparent oversight.

3. Value Capture (Economics)

As in any business model, Cost and Revenue form the Value Capture equation, but in the AI-native model the economics of intelligence create a new logic of profitability by converting learning into margin. This is your Engine of Economics.

• **Costs** shift from human labor (fixed) to compute and intelligence (variable, improving).

• **Revenue** shifts from selling units to capturing a share of the value you create (outcome-based, subscription to intelligence).

• **The Result:** Your margin **compounds** as your system learns. *More use → smarter system → lower cost to deliver → better outcomes → more value to share.*

New paradigm: Profitability comes from continuous learning (intelligence compounding) as you scale because every new customer makes the system smarter and cheaper to operate.

How it applies to Health Platform Case

The conventional economics is based on costs: Personnel (doctors, nurses, admin) and facilities. Revenue: monthly subscription per patient. Result: 15-20% margins, flat with scale.

Cost Transformation: The cost structure lowers human labor and the per-patient costs decrease as the system learns, models improve, and escalations drop.

Revenue Transformation: The revenue model additionally captures share in the value created. Monthly subscription + Outcome-based (15% of healthcare cost savings)

What you're capturing: Not just fees for service, but a share of the value you create. Margins compound with intelligence. As an example: start with investing in learning -33% $\rightarrow$ to improved intelligence at +38% $\rightarrow$ compounding economics at +67% margin.

The AI-native HealthCare Platform

Here's a unified representation of the Health Platform monitoring health signals continuously (wearables, check-ins), detecting early warning signs, and intervening proactively—before you get sick, before you need hospitalization.

Paradigm shift: *Profitability aligned with keeping people healthy, compounding as intelligence improves.*

The Autopsy: MetroBank Case

Let's perform an autopsy on a real business model. Meet "MetroBank"—a regional bank that transformed from a traditional business model into an AI-native intelligence organism.

We'll examine the transformation through the lens of the three engines.

MetroBank as Traditional Model

• **Product:** Standard checking/savings accounts, mortgages, auto loans.

• **Services:** Branch tellers, call centers, mortgage officers.

• **Process:** Loan applications take 5-7 days with manual underwriting. Fraud checks are batch-processed overnight.

• **Cost Structure:** 70% in personnel (branches, call centers, underwriting). 20% in physical infrastructure.

• **Revenue Model:** Interest margins, account fees, loan origination fees.

MetroBank with Micro-autonomy

Value Creation (Intelligence + Compounding Engine)

The Intelligence Engine analyzes real-time cash flow, infers customer intent ("saving for down payment" vs. "emergency fund"), detects spending anomalies to detect fraud or financial stress, and predicts liquidity needs based on historical patterns.

The Compounding Engine learns from every customer interaction and financial outcome, refines predictions, and improves recommendations over time. *These spending categories can be optimized for this goal.*

Example: A customer's account notices an unusual spending spike, cross-references it with historical patterns, infers this is predicted holiday season spending, autonomously moves funds from savings, sends a natural-language notification explaining why. All without the customer opening an app or calling a branch.

What Changed? The checking account is still a checking account. But now it's the delivery vehicle for autonomous financial health management rather than passively storing money.

Value Delivery (Governance Engine)

The Governance Engine personalizes to users' sophistication. Transparent decision-making: *I moved $500 from savings to avoid overdraft in the current account; you can override this or adjust your threshold settings.* Applies deterministic filter *(policy coding)* of business rules, regulatory constraints, and fraud flags. Applies probabilistic check *(risk quantifier)*: high-confidence actions (>90%) execute autonomously; medium-confidence waits for approval; low-confidence escalates to human bankers.

Example: A customer applying for a car loan gets instant analysis: *I can approve $25,000 at 4.2%. Confidence: 92%. Funds in 90 seconds.* When asked *What if I put down $5,000 more?* The system recalculates: *Your rate drops to 3.8% Monthly payment (EMI) decreases by $28.*

What Changed? Service is still delivered by the bank but it is now transformed from tellers and call centers into embedded, adaptive collaboration with each customer, escalating to humans only when needed. 92% of routine decisions are autonomous. Humans handle the 8% that require judgment, exceptions, or complex situations.

Value Capture (Economics Engine)

Cost transformation: Inverted from 70% personnel costs (tellers, loan officers, underwriters, call center) to 30% personnel costs (model stewards, exception handlers, strategic supervisors):

• Fraud detection cost per transaction: $0.15 (human review) → $0.02 (AI detection)

• Loan underwriting cost: $450 (3 days of human) → $5 (90 seconds of compute) per application.

• Customer service cost per interaction: $12 (call center) → $0.50 (autonomous for routine inquiries)

Revenue transformation: Costs and revenue both transformed to enable new revenue models. From fixed fees ($12) to base fee ($8) + micro-earnings on savings generated.

Outcome alignment: Revenue grows when customers' financial health improves.

Value Capture is aligned with the customer's financial health and decreasing costs as the system learns.

MetroBank after Transformation

MetroBank is still a bank. Checking accounts, loans, mortgages, branches—all remain. What changed: micro-autonomy transformed what those products deliver.

- **Checking account:** delivers financial health management.
- **Loans**: with AI-assisted underwriting and custom terms.
- **Bank operations**: run with embedded intelligence that compounds.

MetroBank now works like a **"Financial Resilience Platform"** a living organism that learns and improves with every interaction.

Design Your Intelligence Organism Blueprint

Now it's your turn to design your AI-native business model. Use this framework to map your transformation by reimagining your AI-native business with micro-autonomy.

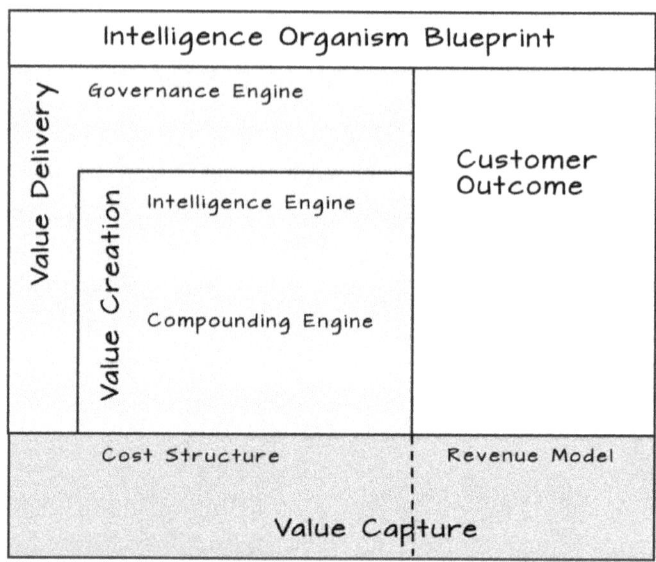

1. Start with Customer Outcome

‣ What result do customers actually want? (Not what you sell them)

‣ What results would customers benefit from that were not possible earlier? *Example: Not "a tractor" but "maximized yield per acre with minimized risk"*

2. Design Value Creation

‣ *Intelligence Engine:* What signals would this system need to sense? What patterns must it recognize? What decisions should it make autonomously?

‣ *Compounding Engine:* What feedback loops will make it smarter? What outcomes need to be captured? How does learning translate to better results?

3. Design Value Delivery

‣ *Governance Engine:* Who are the key human roles? What do they need to see/control to trust the system? What confidence thresholds make sense? When should the system act, suggest, or escalate? How should decisions be explained in natural language?

4. Design Value Capture

‣ *Economics Engine:* What's your new currency (not units, but outcomes)?

How will costs shift from labor to intelligence? How will revenue align with value created?

If you can't sketch this framework for your business, you're still thinking in machines, not organisms.

AI-native Business Models for Physical Products

The AI-native business model of micro-autonomy applies equally to **physical products** like tractors, appliances, machines, etc. The key is to think in terms of outcomes delivered by intelligence, not just the artifact itself. The physical artifact is embedded with sensorized components connected to cloud intelligence and updates over time.

This enables micro-autonomy capability to deliver specific customer outcomes. The business model attaches to this AI capability that makes outcomes observable, measurable, and attributable in the physical world. This enables new economics where a manufacturer may additionally charge for outcomes tied to uptime, efficiency, yield, safety, or other measurable results. In some cases, the physical product may simply become the delivery mechanism for an outcome-based offering.

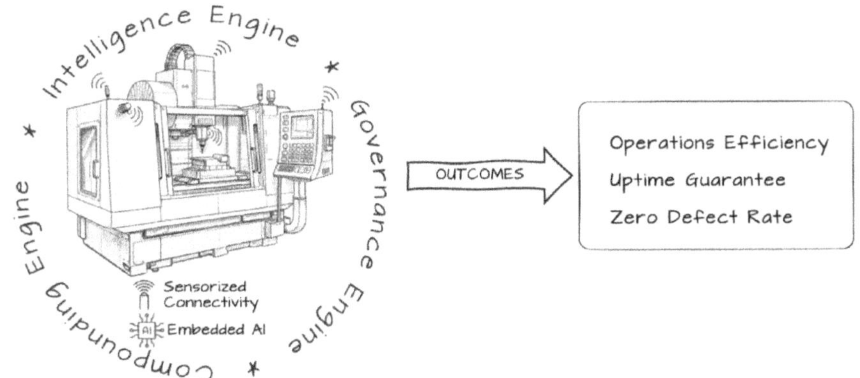

Consider an industrial machine traditionally sold as capital equipment, its performance dependent on operator skill. When sensors are added to monitor vibration, temperature, load, and wear; models detect anomalies, and the machine becomes capable of micro-autonomous decisions. It can adjust operating parameters or flag early failure signals—delivering outcomes such as guaranteed uptime, zero-defect rates, or energy-efficiency. The manufacturer may now price the machine plus an outcome guarantee, or price **only the outcome**, with the machine cost absorbed into the outcome-based model as a delivery vehicle.

Not all physical products require embedded AI, but where it AI can be embedded, the product moves from being a static artifact to a participant in an autonomous and compounding system—one that can support business models that static products cannot sustain.

The core principle is the same: *stop selling tools, start selling intelligent outcomes.*

Build even without Digital Legacy

Consider the infrastructure leap that reshaped telecommunications in countries with poor landline infrastructure. When mobile technology arrived, they didn't modernize landlines, they bypassed them entirely. Landlines became fossils overnight. An entire generation's first and only experience with telephony was mobile—and they never looked back.

The same pattern applies to business domains. The deepest AI-nativity impact will be for sectors that never had digital cores at all. These businesses can create entirely new realities.

Case: Construction Project Management

A mid-sized construction firm doesn't need a "better project management tool." They need autonomous systems that:

• Predict material delays from weather, supplier signals, and historical patterns

• Dynamically reroute labor based on real-time site conditions

• Detect safety violations from drone footage and automatically halt work

• Optimize timelines by learning from every completed project across the industry

• This isn't automation. This is creating operational intelligence where none existed.

Think about it

The past three decades of digital transformation touched only a fraction of the global economy. Sectors like construction, agriculture, logistics, local services depend heavily on legacy tools, spreadsheets, institutional knowledge. These sectors may not have data organized in databases, but they have a lot of accumulated operational data; it's just that they lack the intelligence to harness it.

These sectors that have missed the digital bus are ready for their orbital jump moments.

Unlocking the Orbital Jump

Here's the unlock: Three shifts working together—*scalable reasoning* and *cheap cognition* combined with *knowledge without expertise*—now make possible the initiatives that seemed impossible before.

A regional agricultural cooperative has decades of harvest data, soil samples, and weather records—in spreadsheets and filing cabinets. They couldn't hire data scientists to optimize crop rotation across 500 farms or train their staff to become data experts. Modern AI infrastructure (particularly vector databases) can ingest their existing data—however messy—into intelligent, query-able representations, analyze patterns, and guide their staff through expert-level insights.

The constraints that kept these industries analog weren't technological. They were economic and expertise-related. AI removes those constraints by democratizing expertise they could never afford to build in-house.

The Greenfield Advantage

Interestingly, companies and sectors with the absence of inherited digital process constraints gain an *uncatchable advantage.*

First, the barrier is lower. There is no entrenched workflow to redesign. No legacy software to replace. No migration path to navigate. Only a void to fill with autonomous capability.

Second, the compounding starts immediately. No fighting technical debt of legacy infrastructure. No retraining users on new interfaces. Simply building intelligence-first from day one with no baggage to slow them down.

While incumbents with mature digital infrastructure debate how to integrate AI into existing systems, greenfield players ask, *"If we started today with micro-autonomy as our foundation, what could we build that's impossible for others to replicate?"*

They're not competing with better products—they're building **moats of intelligence** that didn't exist in their industry before.

The ultimate application of AI-native thinking isn't rebuilding the old world smarter but introducing intelligence into sectors that never had it.

That is where unassailable advantage is created. That's the real transformation. That's the real opportunity.

> ☆ ***Street Rule:*** *If your AI strategy is touching workflows that already exist, you're optimizing the present. The exponential opportunity is in enabling workflows that were economically impossible until now.*

CEO Pitch: Vending Machine to Living Garden

"We are not a company with an AI division. We are a live organism of intelligence. Our core reasons, our interface builds trust, and our continuous learning turns into growth. Our competitors are still optimizing their vending machines. We are cultivating a garden that will feed the market for decades."

Up Next

We've transformed the old model and introduced the new one. But how do you actually monetize intelligence? How does this new form of business actually sustain itself? The answer lies in a complete reinvention of business economics, where the old rules of cost, revenue, and profit no longer apply.

Chapter Six

The Economics Engine

When Incentives Flip

Consider a jet engine manufacturer. Traditionally, they sold engines for $15 million each. Airlines bore all maintenance risk. When an engine failed unexpectedly, it cost the airline $500,000+ in downtime, rebooking, and reputation damage, while the manufacturer sold replacement parts at high margins.

The fatal misalignment: *manufacturers profited when engines failed.*

Now imagine inverting the model. You don't sell engines. You sell guaranteed flight hours. The airline pays per hour the engine actually flies. You own the engine, handle all maintenance, bear all failure risk.

Suddenly the model flips. Every failure costs *you* money. Every hour of uptime makes *you* money. Your profitability depends on predicting and preventing failures, not reacting to them.

So you instrument every engine with sensors. You build predictive models. You schedule proactive maintenance. You optimize performance continuously.

As your models improve:

• Engines last longer: *lower replacement costs.*

• Failures decrease: *lower emergency repair costs.*

• Predictions sharpen: *lower preventive maintenance costs.*

Your margin expands as your intelligence improves—while the airline's costs become predictable and their uptime approaches 100%.

This is the **Economics Engine** of AI-native businesses:
- Revenue tied to outcomes.
- Costs that decrease with learning.
- Margins that compound with intelligence.

This inversion is the ultimate expression transforming the economic DNA by aligning it your customer's success. This chapter shows you how to architect this transformation for your business.

The Misalignment of "Selling Stuff"

The industrial revenue model is a simple transaction: you trade a *thing* for money. The value is presumed to be in the artifact. This creates a fundamental misalignment.

• **The seller's incentive:** Move units, protect margins.

• **The buyer's incentive:** Get the promised outcome, avoid hidden costs.

When the *thing* becomes intelligent, adaptive, and integral to the buyer's operations, this transaction breaks down. Charging per seat for an AI that powers your entire go-to-market is like charging per spark plug for an engine. You're pricing the commodity, not the capability.

Your old financial language of COGS, CapEx, per-unit margin obscures your real asset: compounding intelligence.

The Three Shifts of Value Capture

The Economics Engine runs on a new fuel. It requires three simultaneous shifts in how you measure, price, and profit:

1. Shift Your Currency
From Units to Micro-Autonomies

Your fundamental unit of value is no longer a "license" or a "widget." It is a unit of delegated judgment.

• **Bad Currency:** One software seat.

• **Good Currency:** One unit of optimized performance (million micro decisions executed safely).

• **Better Currency:** One point of efficiency gained.

Your pricing should answer: What is the smallest, most valuable packet of autonomous intelligence we can charge for? This is the commercialization of micro-autonomy.

Take the example of Michelin's *Fleet Solutions* that charges by *per mile driven* instead of *per truck tire*. A shift in their fundamental currency.

• **Bad Currency:** One truck tire. *A physical widget with no ongoing value, misaligned incentives.*

• **Good Currency:** One mile of guaranteed tire-performance. *A result of thousands of micro-decisions (inflate now, rotate now, retread now, replace before failure) to optimize tire-performance.*

• **Better Currency:** One percentage point of total fleet operating efficiency gained. *The ultimate packet of autonomous intelligence is the marginal gain in uptime, fuel economy, or safety that the system delivers.*

2. Shift Your Incentives
From Transfer of Risk to Sharing of Upside

A vending machine model (that's a metaphor) transfers all risk to the buyer after the sale. An intelligence model *aligns risk and reward*. You tie your survival to your customer's success.

• **Performance-Based:** Revenue tied directly to value created *30% of the logistics cost savings our AI routing generates.*

• **Outcome-Subscription:** Customer pays for outcome, not tool *$X per month per acre for guaranteed yield increase.*

• **Ecosystem Fee:** Platform earns when users succeed*A small fee on every perfectly matched freelance gig in our network.*

You become a partner, not a vendor. Your success is derivative of your customer's success.

Risk: If your intelligence doesn't deliver, you don't get paid.

Reward: When your intelligence improves (through the Compounding Engine), your margin expands.

3. Shift Your Cost Logic
From Labor to Learning

This is the pivot that makes the economics work. Your largest cost is no longer people *doing* the work. It is the compute, data, and human capital required to *teach* the system that does the work.

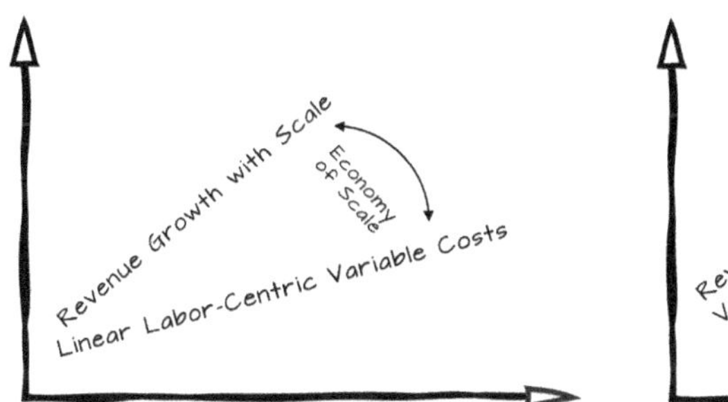

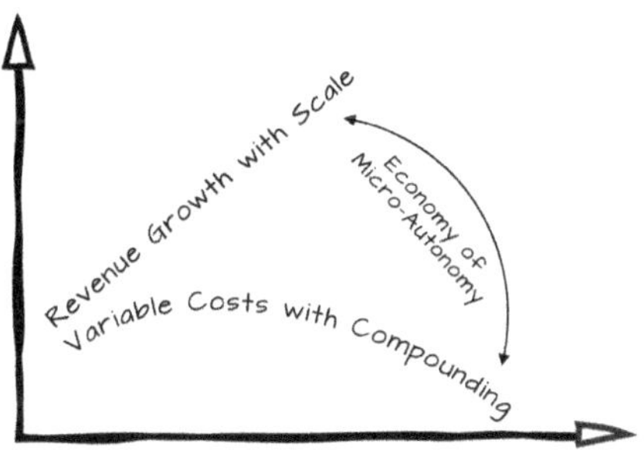

AI-native economics shift from scaling labor to compounding intelligence.

• **The Cost Curve Inverts:** In a traditional model, more scale = more variable labor cost. In an AI-native model, more scale = more data = better models = lower cost per unit of intelligence delivered. The cost structure has a learning curve, not just an efficiency curve.

• **Investment becomes R&D becomes Product:** Every dollar spent improving your Intelligence and Compounding engines is a direct investment in your product and your margin. R&D *is* COGS.

Why Outcome Guarantees are Possible

Outcome-based business models only work when three technical engines operate together. Each engine removes a specific economic constraint. Remove any one, and the model collapses.

Intelligence Engine → Predictability

Enables high-confidence prediction of future states.
Example: Predicts equipment failure two weeks before symptoms appear.

What this unlocks: *"We guarantee 99.9% uptime."*
The business model impact: Without predictability, outcomes are uncertain.
Without certainty, guarantees are reckless.

Governance Engine → Trust

Makes decisions transparent, explainable, and reversible.
Example: Customers see confidence scores, understand reasoning, and retain override control.

What this unlocks: *Customer willingness to accept outcome-based pricing.*
The business model impact: Even accurate predictions fail if customers don't trust them.
Without trust there is no adoption, no conversion of technical capability into revenue.

Compounding Engine → Cost Decline

Improves accuracy and efficiency over time through learning.
Example: Prediction accuracy improves from 75% → 95% over 12 months.

What this unlocks: *Sustained guarantees without margin erosion.*
The business model impact: Guarantees are viable only if costs fall faster than risk accumulates.
Without compounding, guarantees are unprofitable.

Outcome guarantees are not a pricing trick. They are an emergent property of micro-autonomy that makes them inevitable.

The Economic Transition

You cannot flip this switch overnight. You must run a dual-track economy.

Track 1: Legacy Business (Vending Machine)
Continue operating your traditional model. Measure success by traditional metrics: units shipped, revenue growth, margin percentage.

Track 2: Pilot Workflow (Garden—Pick one Function)
Choose one contained workflow in a functional area (say, predictive maintenance for your flagship product). Run it on the new economic model:

• **New Currency:** Value created (dollars of downtime prevented)

• **New Incentive:** Aligned with customer success (base fee + share of savings)

• **New Cost Logic:** Intelligence-driven, not labor-driven

The Bridge: For one or two cycles let these tracks run in parallel.

The pilot's success isn't measured by traditional P&L but by the Value Capture Efficiency (VCE) ratio.

Value Capture Efficiency Ratio (VCE) = Monetizable Outcomes / Cost to Deliver

In plain English: *How much measurable customer outcome do you create compared to what it costs you to deliver it?*

When the pilot shows consistently improving VCE quarter-over-quarter, you've proven the new economics work.

Case: Retainers to Revenue-as-a-Service

A mid-sized growth marketing agency ran on a familiar model of monthly retainers and campaign management fees. Clients paid for activity: ad spend managed, content produced, campaigns launched. When results were strong, relationships renewed. When they weren't, the agency wrote a thoughtful post-mortem and proposed a new strategy.

The fatal misalignment: *the agency got paid whether the pipeline grew or not.*

Now imagine inverting the model.

• **New Currency:** *Conversions and Revenue Uplift.*

• **New Model:** The client pays a base fee plus a share of the measurable revenue impact delivered above baseline through pipeline build-up, conversion rates and revenue generated.

• **New P&L:** Every underperforming campaign costs the agency margin. Every dollar of incremental revenue is shared upside. So, it instruments every channel, builds predictive models, and pushes autonomous micro-decisions—continuously, not in monthly review cycles."

Suddenly the model flips. As their intelligence compounds across clients and campaigns:

• Audience models sharpen *to improve targeting across the portfolio.*

• Attribution becomes precise *to show which decisions drove which revenue.*

• Campaign waste drops *as failing creatives are identified in hours, not weeks.*

Their *margin expands as their intelligence improves*, while the client's growth becomes more predictable and their cost per acquired customer falls.

This is the Economics Engine operating inside a professional services firm. The activity hasn't changed—campaigns, content, channels. But the agency is no longer in the activity business. It is in the revenue-assurance business.

The retainer was the pricing of effort. The new model is the pricing of outcomes—made possible only because the embedded intelligence can now predict, act, and learn faster than any human team reviewing monthly dashboards ever could.

Another Case: Industrial Sensors to Uptime-as-a-Service

A manufacturer of industrial pumps sold sensors and software for a high upfront cost + 20% annual maintenance. Customers bought it, but real value—preventing catastrophic downtime—was hit or miss.

• **New Currency:** *Hours of uptime per quarter.*

• **New Model:** "Uptime-as-a-Service." For a monthly fee per pump, the vendor guarantees 99.9% uptime. Their AI-native system predicts failures and schedules preemptive repairs.

• **New P&L:** The vendor's cost is the intelligent orchestration of parts and field tech. As their prediction model improves, they prevent failures earlier and cheaper. **Their margin expands as they get smarter,** while the customer's operations become seamlessly reliable.

> ☆ *Street Rule: If the ROI of your AI initiative is measured in costs saved rather than value captured, you're building features, not transformation.*

CEO Pitch: The Economics of Intelligence

"Our financial statements track the cost of labor and revenue per unit, while our real expense is the cost of intelligence, and our real product is a probabilistic outcome. We will shift from being a vendor to being an outcome partner. *Our profitability will come from the margin between the value we guarantee and the cost of the intelligence that delivers it.* This is how we will build our Economics Engine that scales with learning, not just with headcount."

Up Next

New economics enable new forms of competition. In this environment, advantage doesn't come from defending a position—it comes from changing the game to accelerate away from everyone else. The winning strategy looks nothing like what's taught in business schools.

Chapter Seven

Competitive Strategy Compounding

The Moat that Fills itself

For decades, Coca-Cola's moat was its secret recipe, locked in an Atlanta vault. For Boeing, it was proprietary metallurgy and patents. For Bloomberg, it was the terminal on every trading desk. These were **static advantages**, fortresses to be defended.

In 2007, Netflix offered a $1 million prize to any team that could improve its recommendation algorithm by 10%. The winning team's code was never used. Why? Because in the two years the contest ran, Netflix's own internal team had already improved the algorithm beyond the winning threshold. The prize wasn't about the code; it was about the global brainpower it attracted to accelerate Netflix's own learning curve.

Today, we see this pattern everywhere. TikTok doesn't just have better content; it has a faster content-understanding feedback loop. Shopify doesn't just offer e-commerce tools; it provides continuous intelligence for conversion optimization. Their real moat isn't content or features—it's their *rate of learning* about what creates value.

This moat doesn't sit still; it deepens with every click, every view, every transaction. Strategy is no longer about building a defensible position. It is about achieving an **uncatchable trajectory**.

Are You still playing "Capture the Flag"?

Traditional strategy is a game of positioning on a static map. You find a hill—a market niche, a patent, a cost advantage—and defend it. You operate on predictable planning cycles, your metrics are financial numbers (lagging, not leading), and your greatest fear is a competitor taking your hill.

AI renders this game obsolete. The map is now fluid, redrawn by intelligence in real-time. The hill you spent five years capturing can be made irrelevant overnight by a competitor whose systems learn a better way to deliver value. A competitor isn't just across the valley; they're playing a different game with a model driven by learning curves, not labor rates. They are competing on time. Their advantage comes from how fast they learn and adapt.

If your strategy document doesn't have a metric for **organizational learning velocity**, it is a recipe for decline. Strategy is no longer a destination. It is a trajectory. Your moat is the steepness of your learning curve from compounding intelligence.

Three levels of Competitive Advantage

Competitive advantage in the age of intelligence compounds across three levels to deliver strategic inevitability.

1. The Loop: Task Velocity

This is where it starts: inside a single functional area or workflow. A learning loop from an action to improve the next action. Micro-autonomy decisions make the process faster, cheaper, and more accurate, creating a local efficiency advantage. Example: Your IT ticket triage is 50% faster than competitors.

The advantage here is fragile. It can be copied or outsourced. It's a better process, not a new paradigm.

2. The Lattice: Organizational Trajectory

Here, micro-autonomy spreads to interlock different functional areas within the organization (Product, Ops, GTM). The learning from one loop supercharges another.

This creates a coherent organizational trajectory. The entire company begins learning in a coordinated direction. Your cost structure bends, and your GTM becomes a learning system. Competitors see you pulling away but can't isolate the cause. This is the stage where micro-autonomy scales from *isolated pockets* to a *systemic capability*.

3. The Gravity: Ecosystem Orbit

This is the stage where your learning loops escape your organizational boundaries to orchestrate partners, customers, and platforms.

You design shared intelligence layers. You plug your learning loops into your suppliers' systems, your customers' operations, into regulatory frameworks. Learning becomes symbiotic, creating collaborative dependency.

Partners succeed when your system learns; customers are more efficient inside your orbit. Competing means not just building a better product but replacing an operating system that improves itself daily. This is the gravity of Amazon's logistics web or Nvidia's AI software ecosystem.

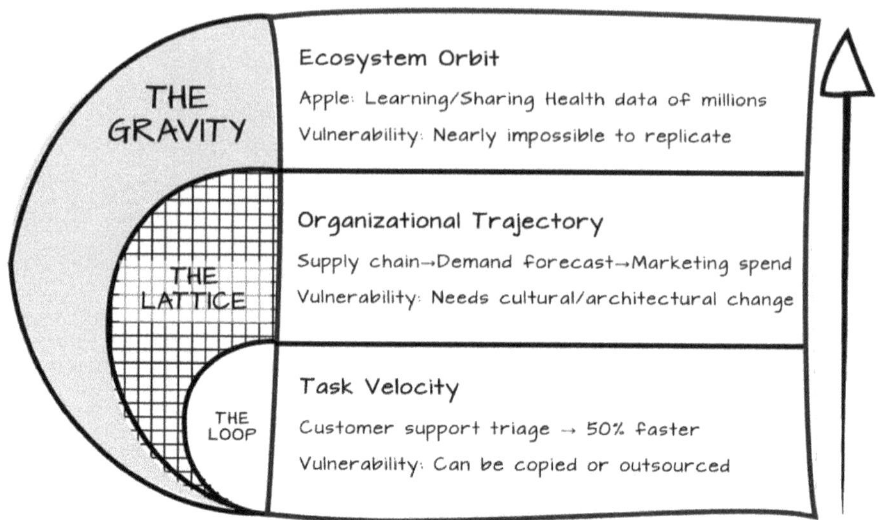

Once intelligence compounds across an ecosystem, competition is no longer firm-versus-firm. It is **system-versus-system**. *New Strategy: Velocity → Trajectory → Orbit.*

This progression—from single loops to organizational lattice, to ecosystem gravity—is the path to uncatchable advantage. *(We'll return to this in Part III with specific moves to navigate it.)*

Now ask, *"What one learning loop, if we mastered, would make our orbital trajectory uncatchable in five years?"* That loop is your new corporate strategy. Allocate capital, talent, and leadership attention to it with ruthless focus. Everything else is a distraction.

Case: Beauty Retailer that stopped selling Makeup

Consider Sephora, a multinational retailer of personal care and beauty products. Its old strategy was about product supremacy: better brands, trendier displays, more stores.

Sephora realized the intelligence isn't in the inventory; it's in the system that **personalizes, recommends, and creates** what each customer actually wants. Its new strategic goal: *Become the central intelligence layer for personal beauty.*

1. Building the Trajectory (Lattice)

▸ Every customer interaction becomes a data node. Purchases, returns, Virtual Artist try-ons, Color IQ skin tone matches, tutorial views—these don't just inform the next sale; they feed a live model of beauty preferences, skin types, and style evolution.

▸ A pattern discovered through its Seoul customers—how certain foundation shades work for specific undertones—refines the system's ability to personalize for similar profiles globally, while the style and color elements enable serving Korean diaspora communities worldwide with culturally adapted beauty recommendations.

▸ Sephora's products don't just get better individually, the entire system's accuracy and efficiency compound.

2. Orchestrating the Inevitability (Gravity)

▸ It partners with beauty brands, sharing its demand forecasts to optimize sustainable production and shade development.

▸ It opens its trend prediction models to emerging beauty creators through an API and developer platform.

▸ It enables third-party beauty services to build on its platform (virtual try-on integrations, skincare regimen optimizers, shade-matching algorithms).

▸ Its "moat" becomes the richness and precision of its **beauty intelligence network**.

Sephora's revenue slowly shifts from selling individual items to beauty subscriptions, outcome-based guarantees (*"find your perfect foundation match or your money back"*), and trend intelligence to beauty brands.

Its strategy is no longer "beat Ulta on store experience." It's *accelerating the intelligence of beauty personalization faster than any other ecosystem.* Competing in retail is now a tactical game. Competing on the learning rate of the ecosystem is the strategic game.

Your new KPIs should measure the *slope of your trajectory* and your *contribution to your ecosystem's learning.*

The Three Signatures of a Compounder

From the Sephora's example, you can recognize an AI-native compounder by three tell-tale signatures:

1. Their rate of improvement increases with scale.
More customers → more signals → better models → increased autonomy → better experience → more customers.

2. Their cost structure flattens as complexity rises.
More products, more brands, more markets—each one *sharpens* the loop; business becomes smoother as it becomes larger.

3. Their strategy becomes self-reinforcing.
Every expansion into new areas, new products, new categories, feeds the core loops and makes the system stronger. *Growth is now an outcome of learning.*

Is your business missing these three signatures?

Why most Companies will never Compound

Most will fail not for lack of ambition, but because their architecture actively prevents learning from accumulating. They fall prey to three structural killers:

1. AI Theater: Intelligence as decoration: dashboards glow, chatbots deploy, demos dazzle. But intelligence remains *performative, not operational.* Nothing accumulates.

2. Governance Paralysis: If intelligence is never permitted to act, it never learns—rigid risk models too may prevent learning. "Human-in-the-loop" becomes *human-as-the-roadblock*.

3. Pilot Addiction: Pilots incapable of consolidating advantage. Fragmented experiments with no shared core. Successes stay local; failures are buried.

These patterns ensure activity without momentum, leaving companies on a flat, linear path while AI-native competitors accelerate away on a compounding curve.

> ☆ *Street Rule: In the old world, you built a moat and sat inside it. In the new world, you dig a channel and watch the intelligence flow. Your strategy is the slope of that channel.*

CEO Pitch: The End of Sustainable Advantage

"There is no more *sustainable competitive advantage*. There is only temporary *advantage sustained by learning*. Our strategy is not a destination. It is the design of our learning engines. We will win not because we have the best plan but because we have the best ability to learn and adapt our plan, minute by minute, at scale."

UpNext

The most powerful forms of intelligence don't respect traditional boundaries between domains or industries. What happens when intelligence flows so freely that it begins to dissolve the walls that have defined competition for decades?

Chapter Eight

The Convergence: Where Industries Collide

The Adjacency Illusion

In 1994, the CEO of Borders bookstore surveyed his empire. He dominated book retail. His nearest competitor, Barnes & Noble, was miles behind. His strategy team mapped adjacencies: music? Maybe. Coffee? They had Starbucks in some stores. The map was clear. They were in the books business.

A startup named Amazon, operating from a garage, had a different map. They weren't in the "books business." They were in the *"infinite selection, frictionless delivery"* business. Books were merely the first, most shippable data packet. Their adjacency map was limitless. They saw a vector, not a territory.

Borders defended its hill. Amazon accelerated along a vector. One is a memory. The other redefined commerce.

This is not the result of diversification but convergence. The moment when intelligence flowing from one domain fundamentally reshapes the economics and possibilities of another.

Today, AI is creating vectors of such velocity and mass that they don't just cross industry lines—they erase them. The winners won't be the best in any category. They will be the **Convergence Giants** whose core intelligence dissolves the walls between categories.

You can spot the company that will dominate a decade by the borders it chooses to ignore. A few will grow by stepping into adjacent industries. But the next AI-native giants will grow by rewiring the boundaries between industries themselves.

From Category to Convergence

What industry are we in? It's the first question at an investor meeting, a strategy session. It provides comfort, clarity, and constraints. But in the age of compounding intelligence, your *category is a cognitive prison*.

Tesla isn't in the *auto industry*. It's in the *sustainable autonomy* business. That vector passes through energy generation, storage, robotics, and AI. Apple isn't in *consumer electronics*. It's in the *intimate computing* business, spanning health, finance, entertainment, and communication.

Strategy in traditional industries is like playing checkers on a fixed industry board. Convergence Giants are playing Go (a strategic board game where you control territory by placing stones) that gain value from their relationships on a grid that spans multiple industries. They don't win by taking your pieces; they win by controlling the flow of intelligence across the entire grid.

The Convergence Vector

Let's dissect a real convergence vector in detail to see how this actually works in practice:

☐ An EV automaker becomes an energy company.
☐ The energy company becomes a robotics company.
☐ The robotics company becomes an autonomy company.
And suddenly it's unclear whether "auto," "energy," or "robotics" were ever the right categories to begin with.

It's not about the car. It's about the cross-domain mesh enabled by its *convergence vector* across *vehicles, energy, and robotics*. No prizes for guessing that company is Tesla.

A Convergence Giant is born when a company identifies and dominates a **Convergence Vector**—a deep, proprietary stream of learning that can be applied to multiple, seemingly disparate domains.

Tesla doesn't see islands but a continuous landscape of interconnected problems. Tesla's core intelligence isn't car-making. It's physical-world perception, prediction, and optimization—a vector that manifests as interlocking flywheels across three domains.

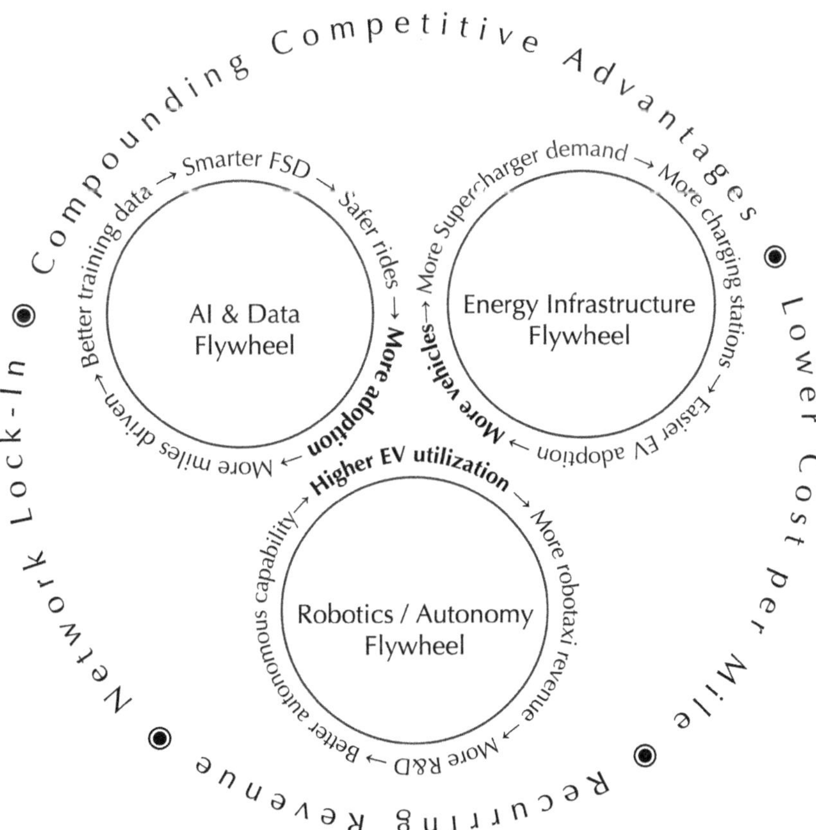

Each domain *feeds the core intelligence*. Car data improves navigation and vehicle AI. Energy data improves grid prediction models. Robotics refines autonomy and real-world interaction.

This is what analyst Mary Meeker calls "super exponential growth," where multiple flywheels accelerate each other. The result: lower cost per mile, recurring robotaxi and energy revenue, and Tesla ecosystem lock-in—which ultimately implies compounding competitive advantage.

Because you are now not competing on cars alone but against a learning system that is also mastering energy, robotics, and autonomy.

The Convergence Flywheel

A traditional EV generates revenue once. A Tesla vehicle generates revenue from:

☐ Car sales
☐ Software subscriptions
☐ Supercharger fees
☐ Smart-grid revenue
☐ Future robotaxi income
☐ Continuous data monetization

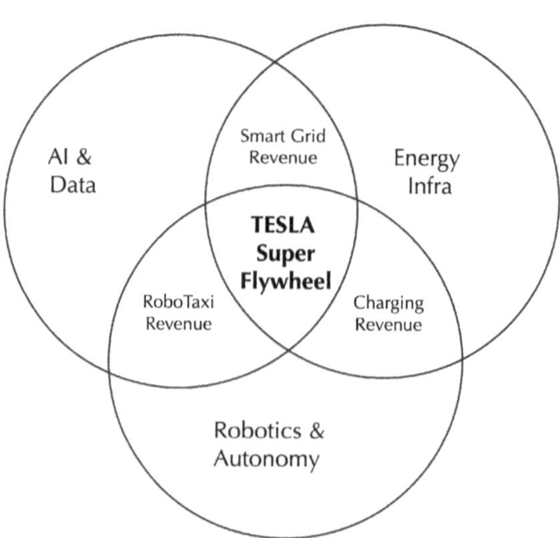

Tesla does not operate separate flywheels for cars, energy, and autonomy. What looks like multiple flywheels is actually one compounding loop, spinning faster as it touches more surfaces of the physical world. Tesla's advantage is not scale alone. It is **learning velocity across connected domains**. Every new deployment strengthens the entire system.

The result? Revenue compounds across streams: hardware sale + software subscription + charging fees + energy services + future autonomy income. The asset (the car) becomes a platform for *lifetime value extraction*. It is the Economics Engine operating at the ecosystem scale.

Convergence Challenge for Legacy Players

Traditional automakers see EVs as electrified cars. Their business models are built for single-sale economics and aftermarket service, not software-driven recurring revenue or energy platforms.

• **Dealer networks** conflict with direct customer relationships and over-the-air value.

• **Service center economics** rely on breakdowns, not predictive uptime.

• **Organizational Silos** (Auto, Energy, Finance) are built to optimize their own P&L, not shared intelligence.

• **Cost Structures** are tied to human-intensive manufacturing and service, not software and compute.

These are not limitations of technology—they are architectural lock-in. Tesla saw the car as the **hardware interface for a scalable autonomy intelligence**—an advantage that cannot be retrofitted.

This Is Not Conglomeration

A conglomerate buys unrelated businesses for diversification. If the new businesses have different intelligence cores, no shared learning between domains, and they don't form an ecosystem bond, you are diverging and not converging. True convergence feels coherent: one DNA that powers different commercial manifestations.

Blueprinting the Next Convergence Giants

We saw the convergence model with a real-life Tesla example. But where will this pattern emerge next? The pattern is repeatable wherever high-friction borders exist between industries that share a fundamental, data-rich problem.

Let's consider two arenas poised for seismic convergence and imagine what they could be.

Convergence Case: A Continuous Care Platform

The healthcare system is a monument to **latency**. A symptom emerges, a patient notices, a doctor evaluates, an insurer approves—each step depends on someone noticing something too late. The solution lies not in making those steps faster, but in *removing the steps altogether* through an AI-native advantage.

The convergence vector? *Predictive, continuous physiology intelligence.*

Devices → Continuous Signals

Wearables, implants, and home diagnostics. Everything that once produced episodic data now produces **continuous streams**. Heart rhythm anomalies, glucose patterns, sleep disruptions—these signals form a living narrative of health.

Healthcare Platforms → Early Interpretation

AI models turn those signals into **early warnings**: *"This pattern resembles early heart failure." "This gait change matches the onset of neuropathy."* Interpretation becomes **proactive**, not reactive.

Insurance → Real-Time Risk Models

Insurers have always priced risk based on **lag**: claims history, demographics, medical records. Now imagine underwriting based on **live physiology**, updated daily. The insurer doesn't just price risk—they help *prevent* it.

The Convergence

A wearable detects a cardiac anomaly, the healthcare platform interprets and triggers a telehealth intervention, the insurance incentives guide care in real time, and the outcomes refine the models.

Across three connected domains—*physiology, insurance risk, and care delivery*—a self-improving decision loop emerges: **health → behavior → risk → care → health.**

The Winner: Not the best device company, insurer, or hospital system. The player who collapses devices, clinical models, and insurance into a single intelligence loop that keeps people healthier at a lower cost. *That is the next healthcare giant.*

Convergence Case: The Infrastructure of Economic Trust

Finance, identity, and compliance are separate industries built on **trust**, which has been slow, manual, expensive, and fragmented. Identity was static. Verification was expensive. Compliance was built around human checks. AI dissolves all of it.

The convergence vector? *Real-time, behavioral trust intelligence.*

The Convergence

Identity becomes a behavioral fingerprint—how someone types, moves, spends, and authenticates, adapting in real time. Finance stops using static scores and instead uses live models updated by behavior: micro-spending shifts, transaction trails, device patterns.

Banks, fintechs, merchants, and platforms begin sharing anonymized fraud signals through federated networks. A fraud attempt in one corner strengthens detection across the entire ecosystem within minutes.

Identity informs Finance → Finance informs Trust → Trust sharpens Identity.

The Winner

Not the bank with the most branches or the fintech with the slickest app. The player who builds the intelligence layer that verifies, underwrites, and protects economic activity across the internet in real time. *That is the next financial giant.*

Plot Your Vector, not Your Portfolio

This is a critical strategic exercise. Gather your team and ask:

1. What is our deepest, proprietary intelligence that we know and others don't?
Example: A logistics company's deep intelligence is "multi-modal optimization under constraint."

2. Where is there friction, waste, and latency that our intelligence could dissolve?□
Example: All of the following suffer from poor "optimization under constraint"—Airline crew scheduling; Hospital resource allocation; Electrical grid load balancing.

3. What would a pilot look like that applied our intelligence to one connected domain?
Example: Offer a hospital an AI-powered "staff and room optimization" service using our logistics core.

If this exercise feels absurd, you're thinking in categories. If it feels thrilling, you've found your vector. Pursuing this, you will become harder to describe. *This is a sign you're on the right path.* The new category is always unclear until it's dominant.

□ *Street Rule: The future doesn't belong to the best in any one industry. It belongs to those who erase the borders between them with a single, compounding intelligence.*

CEO Pitch: The End of Industry Lines

"We are no longer competing with other companies in our sector. We are competing with convergence vectors from other sectors. While we debate market share, someone else is redefining the market around a deeper intelligence—one that sees our domain as a feature of their system. Our goal isn't to win our category. It's to make our category irrelevant by becoming the intelligence layer for a broader convergence."

The Transition

You've seen the new foundations: the living organism that adapts, the economics that transforms, the strategy that compounds to create uncatchable trajectories, and the convergence that redraws industry maps.

But vision without execution is hallucination. The question now is how do we navigate from *What's possible?* to *How do we build it?*

That requires rethinking how your organization operates, assessing whether you're ready to transform, and taking specific actions starting Monday morning. We end not with theory, but with action—Part III is your execution playbook that starts your trajectory.

Part-III: THE PATH TO INEVITABILITY

There's a haunting pattern in business history: companies see the future, understand its implications, invest heavily—and fail anyway.

Kodak was the first to invent the digital camera in 1975. Blockbuster had the capital to buy Netflix for $50 million in 2000. Nokia's executives knew the iPhone would change everything.

They didn't fail because they were blind or lacked capability. They failed because the new physics required a different implementation path than the old physics provided. They had the resources, the talent, the technology but they followed the sunset playbook that had worked for decades.

This is where most transformations die. Not from lack of vision, but from lack of understanding about the new path: organizing around learning instead of execution, assessing readiness before launching, taking focused action to handle tribal (entrenched cultural) resistance.

Part III shows you that path. Competitors are already on it. Here's how you start.

Chapter Nine

The AI-native Organization

The $200 Million Ghost Town

In 2022, a European industrial giant proudly unveiled its "AI Factory": gleaming server racks, massive data screens, a team of brilliant data scientists. The CEO beamed to investors: "This is our future."

Two years later, the factory was a digital ghost town.

The predictive maintenance model could forecast equipment failure with 94% precision. It flagged 17 machines at high risk in Q3. Maintenance supervisors acted on five. Twelve were deferred because shutdown approvals required plant manager sign-off. Two failed. Production lost 36 hours.

The postmortem blamed "model immaturity." But the model was right.

Nothing had changed on the shop floor. Work orders, quality checks, logistics schedules all plodded along exactly as before. The AI lived on dashboards nobody was accountable to use: a glossy wrapper on a hollow core.

The CEO was furious. "We bought the best technology! We hired the best talent! Why isn't it *doing* anything?"

The answer was invisible. *They had built the brain but forgotten the nervous system.* They had intelligence but no architecture to turn it into action.

The bottleneck wasn't the AI—it was the complete absence of human scaffolding required to use it.

Versions of this scenario are playing out in boardrooms right now.

We're Fixating on the Wrong Bottleneck

Every conversation about AI-native transformation obsesses over technology: data pipelines, model selection, compute costs. This is a catastrophic misallocation of attention.

The real bottleneck, the limiting reagent in the entire equation, is not your AI capability. It is your **human scaffolding**: the roles, incentives, cadence, and power structures that surround the technology.

Think of it this way: You can install a fully autonomous robotic arm on a classic assembly line. But if you don't retrain the foreman from overseeing labor to managing system exceptions, the line will stall.

The bottleneck isn't the robot's precision. It's the human system built for hands, not for intelligence. The technology is the easy part. The human operating system is the hard part.

The Human Scaffolding Paradox: *The more powerful and autonomous your AI becomes, the more critical, and difficult, the human architecture around it becomes.*

The New Work is *Human* Work.

Becoming AI-native is not a technology project, it requires an organizational redesign. Your primary task is to construct a new social and decision-making fabric where humans and machines collaborate in a continuous loop of action and learning. This scaffolding has three beams.

Beam 1: The New Roles
From Doers to Deciders.

The loan officer no longer processes 100 applications a day. She designs and tunes the risk model that processes 10,000 and decides the 50 complex exceptions it flags.

Her value shifts from *throughput* to *judgment quality and model stewardship.*

But here is the friction: if she is still measured on volume processed, she will quietly distrust the model. She will override defensively. She will treat AI as interference.

The shift from **doer** to **decider** is not rhetorical. It requires metric redesign—an economic shift. The deciders must be measured on system performance, not task completion.

Beam 2: The New Rhythm
From Meetings to Signals.

The weekly sales forecast meeting is obsolete. In its place is a live *demand signal* channel where the AI alerts the supply team to a surprising spike, and a human inquires.

The AI model sees an unexpected spike in Region B. Marketing, did you launch something? Collaboration happens around interpreting signals and steering models, not reviewing static reports.

But if the weekly two-hour forecast meeting continues unchanged, the signal channel becomes noise. The AI becomes advisory. The old rhythm absorbs the new intelligence and neutralizes it.

Rhythm determines reality. *If cadence does not change, behavior will not change.*

Beam 3: The New Power
From Hierarchy to Permission Slips.

Who gets to change the confidence threshold for an autonomous decision? Who can override an AI-recommended supplier? Who is accountable if an autonomous action goes wrong?

Decision rights must be explicitly recodified not in org charts, but in *digital permissions* embedded in the systems themselves.

If a plant manager must still manually approve every automated recommendation, autonomy is performative. If a junior analyst can override a high-confidence model without logging rationale, governance collapses.

Power in the AI-native firm flows not to rank, but to *accountability for outcomes.* Authority becomes encoded.

Map Your Scaffolding Gaps

Do not look at your technology roadmap. Look at your **human architecture.** Gather your leadership team. Pick one process where you're piloting AI.

Ask three questions:

1. Role Clarity: When this AI makes a mistake, whose job is it to correct it? How are they *measured* on improving the system, not just executing the task?

2. Rhythm Disruption: What meeting or report should this AI make irrelevant? What lightweight ritual should replace it? A daily 15-minute "exception review" instead of a weekly 2-hour "triage meeting"?

3. Power Redistribution: What decision the leader makes in this process could be safely encoded into the system? What decision that a junior employee makes should now require a *reason* to override the AI?

You will find gaps, ambiguities, and resistance. This is the work. The company that solves for this *Human Scaffolding* will outperform the one with a slightly better algorithm, every single time.

The Triad Pod: Your MTU
Minimal Transformation Unit

Don't reorganize the company. Infect it.

Start with a Triad Pod. A minimal, cross-functional team of three people accountable for one stream of micro-autonomy. Each pod *stewards* a single, end-to-end intelligence loop, such as: *Customer Onboarding or Predictive Maintenance or Dynamic Pricing.*

Why three roles?

Because every autonomy loop must do three things simultaneously:

1. Ensure the system's reasoning is accurate enough to act.

2. Earn and maintain human trust.

3. Capture measurable value and compound learning.

Where to Find the Three Roles

Sense Maker
Formerly: Product Manager, Data Scientist, Domain Expert☐
Prime Question: Is our system's reasoning accurate enough to act?
Metric: Decision accuracy, contextual relevance

Trust Architect
Formerly: UX Designer, Ops Lead, Training Manager☐
Prime Question: Do humans trust the system and understand its actions?
Metric: User/operator satisfaction, calibrated override rates

Loop Builder
Formerly: Business Analyst, Process Engineer, Financial Planner□
Prime Question: Is this loop generating measurable value and learning?
Metric: Learning velocity, value capture efficiency

Together they embed autonomy directly inside value streams. They are measured on *Decision Delegation Rate, Trust Score,* and *Learning Velocity,* not task completion.

Capability Gap: Build, Borrow, or Bootstrap?

Now the uncomfortable question: *Do we actually have the right people for the Triad Pod?*

Most organizations assume the bottleneck is AI talent. They conclude they must hire aggressively before starting. That instinct is often wrong.

The real gap is role clarity, cross-functional translation, and incentive alignment. So audit before you recruit.

Bootstrap/Build

- Is the gap technical fluency? *Train domain experts in AI literacy.*

- Is it systems thinking? *Pair technical staff with business leads.*

- Is it the authority to act? *Rotate high-potential operators into pod roles.*

- Is it incentive misalignment? *Change the metrics.*

Bootstrap is often the fastest path to clarity. It is slower and imperfect at first, but this *builds institutional intelligence* from operating the loop. Choose this when autonomy is strategic to your competitive advantage.

Borrow

Embed external experts temporarily.

☐ Fractional AI strategists
☐ Data science partners
☐ Specialized model builders

Their mandate should be explicit: deliver output and transfer capability. Borrowing is a bridge, not a substitute for ownership. If the pod becomes dependent on outsiders to function, autonomy is not institutionalized—it is rented.

The Strategic Choice

The real decision is not Build vs. Borrow.

• If autonomy is core to your advantage, you must build institutional muscle.

• If it is only supportive, borrowing may be sufficient.

Either way, do not delay the first pod waiting for ideal conditions.

The Inevitable Resistance

Your biggest hurdle won't be technology—it will be unconscious sabotage by information *gatekeepers* who fear obsolescence. The manager who hoards decision-making authority, the expert who protects proprietary knowledge, the executive who equates oversight with value.

Your job: Move their power from *gatekeeping to sense-making.*

How:

1. Frame the shift: "Your expertise is too valuable to spend on routine decisions. We need you designing the systems that handle them."

2. Measure differently: Shift metrics from *How many decisions do you make?* to *How well does your system make decisions without you?*

3. Create new status symbols: Celebrate the leader whose system runs autonomously with low override rates.

Note: People don't resist change; they resist loss. Show them what they gain. That identity shift is the transformation.

Case: The Loyalty Pod

A telecom company launches a Triad Pod to *reduce voluntary churn.*

The Sense Maker builds a model predicting churn risk from call logs, usage, and competitor offers.

The Trust Architect designs the retention agent interface to show the risk score, key reasons, and AI-suggested "save" offers in plain language.

The Loop Builder instruments the compounding loop: what offer was made, was it accepted, did the customer still churn? They track how quickly the model improves.

Within a quarter, this pod should *own* the churn metric. Not executing a retention *process* but systematically learning *how to maximize lifetime value.* They become the company's foremost experts on customer loyalty—not the marketing department, not the call center.

> □ **Street Rule:** *If your AI initiative has a technical roadmap but no organizational redesign plan, you're building a museum exhibit, not a transformation. The org chart changes first, or the technology never leaves the lab.*

CEO Pitch: The Human Architecture Mandate

"Our greatest investment now is not in more GPUs. It is in redesigning how our people work, think, and wield authority. We are not building an AI company. We are building a company where brilliant humans are freed from routine judgment to focus on strategic judgment. The AI handles the predictable. Our people handle the exceptional. Our competitive advantage will be the quality of our human scaffolding."

Up Next

Before you can build, you must assess the ground you're building on. Many transformations fail not from lack of vision, but from starting before they're structurally ready. How do you know if your organization is prepared to make this leap—and how do you get it ready?

Chapter Ten

The Demo-to-Deployment Gap

From Magic to Mechanics

Every leader has seen an AI demo that feels like magic. An agent that writes a report in seconds. A chatbot that answers complex questions. A system that codes from a description.

The natural question is, "When *can we deploy this?*" Too often, the answer is "*Not yet. It's not ready for business.*"

What changes when you move from **demo** to **deployment** and why most companies get stuck in between is the focus of this chapter. This gap isn't technical; it's operational discipline for *production readiness.*

Areas of Production Readiness

A demo has to work once, impressively. If it fails, you reset and try again. A production system must work reliably, every time, at scale—with accountability, safety, and clear rules of engagement.

So, what does this entail?

Trust Engineering

Confidence Scoring

Knowing When to Act vs. Ask

Best Practice: System should know its own confidence score to **Act** (>90%), **Suggest** (70-90%) or **Escalate** (<70%). This is a required layer of business logic.

Validation

Hard Rules to Override AI Confidence Best Practice: Always run every AI suggestion through business rules. For example, regardless of AI confidence, block any loan approval that exceeds regulatory debt-to-income limits. Business rules always trump probabilistic confidence.

Risk-Calibrated Thresholds

One Size does not Fit All Best Practice: Match thresholds to business risk and consequences. Routine email routing may need 80% certainty, while a medical recommendation or large-loan approval should demand 98%+.

Accountability Architecture

Audit Trails

Every Decision must be Explainable and Auditable Best Practice: Log everything: prompt, reasoning, confidence, rules checked, human approvals, and outcomes. It is your evidence for regulators, auditors, and liability protection.

Human-in-the-Loop

Escalation as a Feature Best Practice: Ensure seamless workflows hand-offs that give humans full context, suggested actions, and a simple approve/override interface when confidence is low or stakes are high.

Understanding Infrastructure

Explainability

"Why" matters as much as "What" Best Practice: Ensure the system can answer "Why?" in business terms for every decision. "Loan denied because debt-to-income ratio exceeds 40% and three recent credit inquiries suggest financial stress."

Many AI agent platforms demo orchestration but business-critical decisions require **governance**—ask *"How do we configure this for our unique business risks?"* It is the **Governance Engine** that wraps the AI's intelligence with the safety, rules, and accountability essential for responsible real-world deployments. And that's non-negotiable.

The Path Across the Gap

Companies like **Stripe** (Radar fraud detection) and **Tesla** (Autopilot) didn't start with mature systems. They started with one narrow use case, proved the value, and scaled the architecture. They built them from the ground up over years, with dedicated engineering teams and massive data volumes. You can too.

Your crossing begins the same way: pick one contained, high-volume, lower-risk workflow—a process loop or service loop—the repetitive, measurable workflows where AI can prove value quickly.

• **For the Intelligence Engine:** Use hosted LLMs, don't build your own models.

• **Stand up a lightweight Governance Engine:** Start with simple confidence thresholds and a few business rules.

• **Integrate with your existing systems:** The AI layer sits on top of your ERP, CRM, or ticketing system—it doesn't replace them.

• **Compounding Engine:** Log outcomes and learn. Capture decisions, confidence scores, and results. Review weekly. Adjust thresholds manually at first, scale gradually.

• **Build the Business Case:** Use the results of the pilot/demo to project the full-scale impact.

□ *Cost Reduction:* Hours saved, errors eliminated, reduction in operating expenses.□

□ *Revenue Growth:* New revenue models, LTV increase, market share growth.□

□ *Productivity Gains:* Process efficiency, quality improvement, faster response time.

• **Expand once the pattern is proven:** Move to adjacent functional areas, add more rules, and begin automating the learning loop.

The ROI Progression

Many organizations encounter AI first through operational use cases—agents that handle tickets, copilots that assist employees, or workflows that remove human effort from repetitive tasks. These are valid entry points, as they make the mechanics of micro-autonomy tangible.

While they deliver productivity gains, ROI is hard to quantify. Once the readiness DNA is established, quickly move to engage where decisions materially change results—and where those results can justify the investment.

Phase 1: Operational AI

• **What:** Automating tasks, assisting employees

• **Example:** AI approves invoices automatically → productivity gain

• **ROI Claim:** "20% faster processing"

• **Problem:** Hard to quantify, easy to dismiss as "nice to have"

Phase 2: Strategic AI

• **What:** Changing business outcomes

• **Example:** AI predicts which customers will churn and recommends retention strategies

• **ROI Claim:** "Reduced churn by 15%, saving $4M annually"

• **Advantage:** Quantifiable, business-critical, justifies investment

Phase 3: Transformational AI

• **What:** Enabling new business models

• **Example:** AI enables outcome-based pricing and guarantees

• **ROI Claim:** "Increased customer lifetime value by 40%, created new revenue stream"

• **Result:** Changes how you compete, not just how you operate

To Summarize:

☐ Operational AI replicates workflows.
☐ Strategic AI improves business outcomes.
☐ Transformational AI changes the logic of the business.

The Bottom Line for Leaders

The demo-to-deployment gap isn't a technical problem; without **governance** the most advanced AI model is just a demo.

Measure success in business terms: dollars saved, revenue generated, productivity improved.

Vanguard is one of the world's largest investment companies with 30 million investors. They didn't celebrate 95% model accuracy—they celebrated

$500M in ROI and 25% productivity gains for developers. That's the language of transformation.

The winners in the AI-native era won't be the ones with the smartest algorithms—they'll be the ones who turn AI from a novelty into a competitive engine.

Now, let's turn to execution.

Three Nudges Before You Execute

1. Start small, then scale. Avoid building elaborate infrastructure or hiring new teams for your first micro-autonomy loop.

2. Build parallel, don't replace. Run new systems alongside legacy ones, move workloads gradually as trust builds, and sunset the old.

3. Deploy in shadow mode first. Where *system advises but doesn't act.* As accuracy improves, enable autonomous actions and gradually reduce human review to exceptions.

Note: Lacking digital infra ≠ lacking data. Modern AI can process data from spreadsheets, PDFs, and legacy tools—enough to begin your AI-native journey.

> □ ***Street Rule:*** *A demo proves capability. Deployment demands readiness. That bridge is built with culture, governance, and ownership.*

CEO Pitch: From Demo to Business

"We've seen what AI can do in demos. Now we must build what it will do in production. That means engineering trust, not just intelligence—confidence scoring, audit trails, human escalation, and business rules that override AI confidence when needed.

Our measure of success shifts from 'Does it wow?' to 'Does it work reliably, safely, and accountably?' The gap between demo and deployment is where most AI initiatives die. We will cross it by building governance first, then scaling intelligence."

Up Next

All of this leads to a single Monday-morning question: what do you actually *do*? Theory becomes action in six deliberate moves—starting not with a grand plan, but with a single, closed loop that begins building momentum from day one.

Chapter Eleven

The Six Moves to an AI-Native Trajectory

Your Trajectory Starts Now

It is 8:47 AM on a Monday. You have finished this book. The ideas are fresh, the conviction is high. You walk into your office. What do you do? Form an AI strategy committee? Ask your CTO for a capabilities assessment? That's old physics of building monuments to scarcity.

The trajectory doesn't start with a plan. It starts with a **move**. Strategy in the age of AI is not a document. It is the direction of motion. Your job as a leader is to create that first, irreversible vector.

The Six Moves

Here are the Six Moves designed to build momentum from day one. This is not a checklist. It is a sequence. Each creating a momentum that makes the next inevitable.

MOVE 1: The Strategy—*Choose Your Game*

Inside your leadership team, ask:

- *Are we using AI to improve our current operating model?*

• Are we creating capabilities that couldn't exist before? That our competitors cannot match?

Don't limit your search to digitally mature processes. Often the best candidates are the ones with fewer entrenched workflows to defend. Both are valid. But they require different approaches.

The Optimization Play

Take an existing process and embed it with intelligence. Reduce fraud losses by 15%. Improve inventory forecasting. Increase sales routing precision. This is lower risk and easier to measure. Start here for early wins to build momentum.

The Void Play

Find functional areas that were too variable, too judgment-heavy, or too expensive to systematize, and as a result their workflows never existed. Offer instant micro-insurance products previously impossible to underwrite Create real-time dynamic pricing in a market that historically used static contracts. The Void Play can create a durable advantage in a territory that was previously empty.

> *Optimization builds efficiency.*
> *Void builds asymmetry.*

Either way, choose deliberately, know which game you're playing before you pick your next move.

MOVE 2: The Diagnosis—*Declare Your Beachhead*

Don't look for "AI use cases," but a stream of micro-decisions where *better judgment* creates disproportionate value and where that value can be measured to generate learning.

☐ *In a hospital:* Triage, diagnosis, bed allocation, supply ordering, appointment scheduling☐

☐ *In a retailer:* Demand forecasting, dynamic pricing, markdown optimization, staff scheduling.☐

☐ *In a factory:* Predictive maintenance, quality inspection, energy optimization, supplier risk☐

☐ *In fintech:* Fraud detection, loan triage, portfolio rebalancing, compliance checking

Ask: *Where do we make high-volume, repetitive judgments today? Where does performance vary wildly based on individual experience? Where do we feel blind or constrained?*

Evaluate beachhead candidates across three dimensions:

• **Business impact** (quantifiable: cost reduction, revenue lift, productivity gain)

• **Technical feasibility** (buildable with today's tools and existing systems)

• **Data sufficiency** (enough historical and ongoing data to train intelligence)

Choose where all three align.

Rule of thumb for Beachheads

Good starting workflows = Medium business impact + simple implementation+ low operational or regulatory risk. Such functional areas create fast learning and allow micro-autonomy to prove itself and earn trust.

Bad starting workflows = High regulatory + high reputational + high downside concentration. Even when technically feasible, they are poor places to begin.

Your first beachhead is not where failure would be most costly, but one where learning would be most forgiving.

MOVE 3: The Genesis—*Close the First Loop*

From your beachhead candidates, choose **one** and publicly commit to closing a single, complete learning loop around it. This is your **Genesis Loop**—small enough to feel almost trivial, yet fully closed so it can learn and improve.

A close loop has no open ends. No human "review" that doesn't feedback as data into the system.

Before you begin, confront one question: *Can this loop actually close?*

• If outcomes cannot be measured, *the model cannot improve.*

• If the system can predict but not act, *you are running a demo.*

• If ownership of decision quality is unclear, *no one will steward it.*

• If early imperfection is intolerable, *learning will be shut down at the first error.*

You do not need enterprise-wide transformation to start. You need sufficiency in this one functional area or one workflow—enough data, clear ownership, and tolerance for supervised autonomy—to allow the loop to operate and learn.

If you assign the ownership of this loop to IT, or CTO without P&L ownership, it will become tech upgrade and not a business transformation project. The loop must belong to the business.

This is how a closed loop should operate:

Signal → Model (Intelligence) → Action → Measured Outcome → Improved Model.

Example Loops:

• Retail Replenishment: Store sales data → AI forecast → automatic re-order → measure stockouts or excess → improve forecasting.

• Industrial Maintenance: Machine sensor data → AI predicts failure → schedules inspection → technician confirms/denies → model improves.

• Customer Support Triage: Support ticket text → AI routes to right team → track resolution time & customer satisfaction → improve routing accuracy.

A tiny but closed loop teaches far more than a grand, half-finished experiment.

MOVE 4: The Breakthrough—*Remove the Friction That Blocks Learning*

Your genesis loop will quickly reveal organizational friction: manual approvals continue, data availability requires copy-paste across disconnected systems, misaligned KPIs favoring old process.

These do not prevent the loop from existing. They prevent it from compounding. Your job is not to work around these walls, but to deliberately dismantle them.

Identify Friction:

• Where do signals get stuck in emails or spreadsheets?

• Where do decisions require meetings or reviews that add no unique judgment?

• Where the system overrides are unlogged and invisible.

• Where is the user behavior defensive to prevent learning.

Empower your team to architect **learning flow**. Give them the authority to:

1. Build simple data pipelines,

2. Replace manual approvals with an automated, auditable checkpoint,

3. Log every override as training data, and

4. Shift metrics from "tasks completed" to "learning signals captured" or "autonomous decisions made."

You are no longer optimizing mere efficiency. You are optimizing **velocity of insight**. Technology is rarely the real barrier—organizational inertia is. Removing these frictions makes the transformation irreversible.

If the loop does not shorten its learning cycle over time, you are running static automation. When you see acceleration it signals that autonomy is becoming structural.

MOVE 5: The New DNA—*Re-architect the Operating Model for Supervised Autonomy*

As your loop matures, formalize the redesign of roles and responsibilities around the new supervised autonomy.

Assemble a small core team—your Triad Pod of Sense Makers, Trust Architects, and Loop Builders.

• Redesign the work to flip the ratio of human vs. autonomous decisions.

• Move humans from executing to supervising, training, and handling exceptions.

• Measure success by the expanding scope and reliability of micro-autonomy under their stewardship.

Celebrate the emergence of a new operating rhythm: brief daily syncs focused on autonomous decisions made, exceptions handled, and lessons learned.

Caution: If your people are still gathering data manually or processing routine transactions, you have merely automated; you have not truly redesigned for AI.

MOVE 6: The Multiplier—*Connect Across Functional Areas*

Once you have a few strong, closed loops running smoothly, connect them to adjacent or connected business areas within the enterprise. Let intelligence compound as it flows across these functional areas, activating an organization-wide Learning & Value Engine.

Examples of connection:

• Retail: Link demand forecasting to logistics optimization—real-time forecast shifts automatically reroute deliveries.

• Healthcare: Connect patient triage insights to staff scheduling—predicted influx pre-adjusts shifts.

• Manufacturing: Tie quality inspection findings to procurement—defect patterns trigger supplier reviews.

This is where true, uncopyable advantage emerges. The **lattice loop** becomes your deep moat.

Competitors may spot individual loops, but they cannot replicate the accumulated learning that now flows uniquely across your connected functional areas.

Measure ongoing progress by learning velocity: time-to-insight, model improvement rate, percentage of decisions handled autonomously, and rising complexity levels of those decisions.

Where Next: From Lattice to Gravity

The Six Moves take you from your first "Loop" to a connected "Lattice" of learning within your organization. This is where you should focus in the first 12-18 months.

Once your Lattice is strong and learning velocity is compounding, the **next logical evolution** is to extend intelligence beyond your walls: to suppliers, partners, customers—that's **Gravity** (Chapter 7). And finally beyond your domain into adjacent industries for **Convergence** advantage (Chapter 8).

First, build your lattice. Make learning reliable, measurable, and scalable inside. Because you can't orchestrate an ecosystem before you've mastered your own organism. Then, and only then, will you have the coherence and confidence to invite the outside world into your intelligence orbit.

New Economics Unlocked

The Six Moves build your trajectory. But trajectory alone isn't the goal—transformation is. And transformation reveals itself in economics.

As variability drops and accuracy of judgment improves, new economics becomes possible:

As your learning loops mature—variability drops and judgment accuracy improves—new economics becomes possible:

• **Performance-based pricing**: Guaranteed outcomes because they are now predictable

• **Reduced reserves:** Buffer inventory and contingency costs collapse as uncertainty vanishes

• **Guaranteed SLAs:** Commitments become enforceable because the system self-corrects

• **Dynamic contracts:** Price adjusts in real-time based on value delivered, not effort expended

The specific levers depend on your industry and business model. But the pattern is universal: as intelligence compounds, costs decrease, outcomes become predictable, and economics that were previously unviable become inevitable.

This is when the AI-native advantage becomes visible in margins.

The Six Moves build the architecture. The new Economics Engine is what that architecture enables.

□ ***Street Rule****: Strategy in the age of AI isn't a document. It's a direction of motion. Pick your first loop. Close it. Then you'll know what the second one is.*

CEO Pitch: Direction of Motion

"Our transformation doesn't start with a five-year plan. It starts with six moves—beginning Monday morning. We will declare one beachhead, close one learning loop, remove one friction point, redesign one team around stewardship, then connect to one adjacent functional area.

We are not planning a destination. We are starting a trajectory. One closed loop becomes two. Two become a lattice. The lattice becomes gravity—connecting across the ecosystem. Our strategy is now a direction of motion—and it starts with move one."

The Final CEO Question

At the start of this book, you might have asked: *"Where should we use AI?"* That was the wrong question.

You now know the right question: *"Where will intelligence accumulate in our industry, and how do we position ourselves in its flow?"* That is the question your board should debate.

But here's the Monday morning version: *"What's the one closed loop we can complete in 90 days that will teach us more about that flow than any strategy document ever could?"* That's the question your team should debate today.

Your trajectory starts now. Not with a big bang, but with a single, closed loop. **Build it.**

Epilogue

The Shift That Will Define a Generation

If you strip away the models, the data, the hype cycles, and the jargon, you are left with a simple, seismic truth: *the world is rearranging itself around intelligence.*

Not artificial intelligence—that's a limiting term for what is happening. It is embedded intelligence—emergent and alive, flowing through systems that learn from every action, every decision, every outcome. For centuries, we built businesses that depended on people noticing patterns. We are now entering an era where businesses will act on patterns long before any human can name them.

That difference may sound subtle. It is not.

It is the difference between:

☐ Adapting and compounding
☐ Responding and anticipating
☐ Surviving disruption and setting its pace

To the Founder

This is your moment of historic asymmetry. You carry no legacy systems, have no organizational inertia, and need no permission to build what has never existed. Your blank canvas is both a privilege and a weapon. Use it to outmaneuver companies a hundred times your size; all you need is logic.

To the Enterprise Leader

This is your moment to rise, your moment of leverage. You have scale, customers, data, and history. What you need now isn't more technology, but a new operating model that turns those assets into momentum. Your greatest risk is your installed thinking, get over it.

To the Investor

This is your moment to see clearly, your lens correction moment. Look past the demos and the valuations. Look for the *loops*. Look for the companies that have closed a feedback circuit, however small. Look for leaders who measure learning velocity, not just quarterly growth. The winners will not be those with the most AI patents, but those with the tightest, fastest cycles of intelligence.

To an Individual, in any role

This is your moment to redefine your worth, to recalibrate your career. Your value will no longer be what you know, but how well you learn—and how well you teach the systems around you. The most sought-after skill will be *judgment literacy*: the ability to interpret uncertainty, to supervise intelligence, to handle the exception that defines the rule.

This book was not written to give you the answers.

It was written to change the questions you ask.

You will no longer ask, *"How do we do this faster?"*
You will ask, *"Why are we still doing this at all?"*

You will stop asking, *"What features should we build?"*
You will start asking, *"What should our system learn next?"*

You will abandon the question, *"How do we protect our advantage?"*
You will embrace the question, *"How do we steepen our learning curve?"*

No industry is immune. No company is safe.
But every company is capable—if it chooses the right trajectory.

The transformation described in these pages is not a prophecy. It is already underway—in the labs, in the pilot projects, in the quiet rewiring of processes in companies you compete with every day. The physics has changed. The ground has shifted. You are already standing on the new terrain. The choice is not whether but how you will build on it.

The next decade will not be written by those who adopted AI. It will be written by those who understood what AI makes possible.

The shift is not coming. The shift has begun.

And like all great transformations do, it starts not with a grand plan but with a single, closed loop.

Build yours!

Appendix A: The Startup Builder's Playbook

7 First Principles of AI-native Building

This appendix is for founders, product leaders, and builders ready to starti from zero.

While the core book addresses transformation for organizations of any size, these seven principles are your tactical blueprint for building a company where intelligence is not a feature but the foundation.

Each principle is both a strategic reframe and a tactical imperative. Together, they form a new blueprint for building startups where AI is not a bolt-on but the backbone.

Note: *Avoid the temptation to build "AI-powered X." Your advantage isn't about being better than incumbents, but building with intelligence as the foundation that is impossible for them to replicate.*

Phase 1: Ideation & Foundation

Principle 1. Intelligence Is the Product

Don't add AI to a product. Make intelligence the product.

Ask: Is intelligence itself your product? Or are you adding AI to something?

AI-native ventures don't "apply AI" after building software. They deliver learning, judgment, adaptation, and automation as the core value.

What it rewires: Value Proposition. You are selling judgment, not tools.

Example: GitHub Copilot doesn't *contain AI*; it is *AI in action*.

Implication: Founders must reimagine what they're selling—not features, but intelligence. Your Intelligence Engine isn't a tool; it's a reasoning partner.

Principle 2. Data Without Domain Is Dangerous

Vertical advantage = AI + domain wiring.

Ask: Have you picked a specific vertical? Do you understand its domain logic deeply?

Generic AI models are not defensible. Contextualization is. Embed deep domain logic—workflows, edge cases, norms—into the AI loop.

What it rewires: Defensibility. Context > Models; Verticals matter.

Example: Legal AI startup isn't valuable because of GPT; it's valuable because it understands contract nuance.

Implication: Pick a vertical. Wire the domain into the AI logic. Your Intelligence Engine needs domain-specific training.

Principle 3. Learn Before You Launch

Your MVP is not code. It's comprehension.

Ask: What can you learn before you build? What existing data/traces can train your system?

AI-native founders validate not by launching first, but by training intelligence before launch. Use existing user traces, feedback loops, simulations, small datasets to refine intent and accuracy.

What it rewires: MVP Strategy. You train before your ship.

Example: Agent-based CRM that learns user preferences through email history before ever replacing anything.

Implication: Pre-launch phase is now a training phase. Your Intelligence Engine starts training before you have customers. Your Compounding Engine activates when real feedback begins flowing.

Phase 2: Design & Architecture

Principle 4. Agentic First, App Later

Start with agents. Don't default to apps.

Ask: Are you starting with agents or apps? Is your default interface conversational?

Traditional thinking says: *Idea → App → UX → GTM.*
AI-native flips this: *Intent → Agent → Interface-as-needed.*

Interface follows capability, not convention. UI may be voice, chat, workflow, or none at all.

What it rewires: UX and Delivery. No app necessary; the agent is the interface.

Implication: The "product" may be invisible. That's power. Your Governance Engine might be conversational before it's visual.

Principle 5. Compute Abundance, Human Scarcity

Design for abundant cognition, scarce attention.

Ask: Have you mapped which tasks are "compute" (AI) vs. "attention" (human)?

Founders must shift from "resource scarcity" to abundance thinking, where reasoning, summarization, perception = near-zero cost. Use AI for heavy lifting. Conserve human energy for judgment, emotion, ethics.

Think: Agents do; humans decide.

What it rewires: System design. Conserve human bandwidth, not compute.

Implication: Product architecture must separate *thinking tasks* from *feeling or deciding tasks*. Your Governance Engine should optimize for human attention, not human computation.

Principle 6. Every User Is a System

Design for the individual, not just the role.

Ask: How will your system adapt to each individual user from day one?

AI-native systems personalize at user granularity, not persona level. They infer context, intent, preferences, and evolve with the user.

Think: Each user has a unique operating system and your product adapts to it.

What it rewires: Personalization. Design for one, not segments.

Implication: Build for per-user memory, not just shared dashboards. This is the Governance Engine at work—personalizing thresholds, routing, and collaboration.

Phase 3: Launch & Scale

Principle 7. Strategy Is a Learning Loop

GTM is now a compounding loop.

Ask: Is your GTM designed as a learning loop? How does each customer improve the system?

AI-native companies don't launch and iterate. They deploy and evolve. Every customer interaction becomes training data. Your moat is no longer patents—it's compounding contextual learning.

What it rewires: GTM and moat. Feedback = Flywheel = Fortress.

Implication: GTM = *Learn* → *Adapt* → *Lock-in* → *Expand*. Product improves with use. This is your Compounding Engine in market.

The Pivot: From Traditional to AI-native

Many founders begin with a traditional idea, then realize AI changes everything. Here's how to know when to pivot:

Signs you should pivot to AI-native:

1. Your "secret sauce" is really pattern recognition or judgment

2. You're hiring experts faster than you can scale

3. Customer outcomes vary widely based on who serves them

4. You're building dashboards instead of decisions

How to pivot without starting over:

1. Identify the core judgment in your current model

2. Build one closed loop around that judgment

3. Measure improvement velocity (not just accuracy)

4. Gradually expand autonomy as confidence grows

Example: A traditional tutoring platform pivots to AI-native when it realizes: *Our value isn't connecting tutors to students; it's knowing which teaching method works for each learner.*

Case: Building an AI-native Legal Startup

Old Approach

"We'll build a legal research platform with AI features."

AI-native Approach (Using 7 Principles)

1. Intelligence Is the Product: Selling "Contract Risk Assessment as a Service."

2. Domain Wiring: Former M&A lawyers teach the model industry nuance.

3. Learn First: Train on 50K anonymized contracts before writing a line of code.

4. Agentic First: Email service: "Forward contract to contracts@ for risk analysis in 5 minutes."

5. Compute Abundance: AI analyzes clauses; lawyers focus on client relationships.

6. Every User Is a System: Learns each lawyer's risk tolerance and preferred style.

7. Strategy Is a Learning Loop: Every analyzed contract makes our system better.

Result: Not a better legal research tool, but a **legal intelligence partner** that becomes indispensable.

The Founder's Mindset Shift

Building AI-native requires a fundamental shift in how you think about your company:

Traditional Startup Thinking	AI-native Startup Thinking
Build features users want	Build intelligence that learns what users need
Scale through headcount	Scale through learning velocity
Protect with patents	Protect with proprietary feedback loops
Optimize for user acquisition	Optimize for learning per user
Product roadmap = feature list	Learning trajectory = capability expansion
Pivot when features don't work	Pivot when learning plateaus

The startup that understands this will outpace competitors who see AI as just another technology to adopt. They're not playing the same game. They're building a different kind of company entirely.

Final Word to Builders

Remember: you're not just building a product. You're cultivating intelligence.

The companies that will define the next decade aren't those with the most funding or the flashiest demos. They're the ones that close their first learning loop, then their second, then connect them into a compounding system.

Start small. Start closed. Start learning.

Build your first engine. Then watch what becomes possible when the three engines turn together.